TOW

Do *you*
the wo ... ess of
motherh ... ness of sex, the
intoxication of national patriotism, the itch for infinite freedom, and the respect for industry, in favour of man-centred population restriction involving sex for fun and implying world government, managed ecology, and education for leisure?
– And do it all before the twentieth century has run out?
We don't have to, you know.
It's just that if we don't, our civilization will be destroyed in thirty years, that's all.

Also by the same author, and available in Coronet Books:

TODAY AND TOMORROW
(part 1 of Towards Tomorrow)

The Tragedy of The Moon
Asimov on Astronomy

Towards Tomorrow

Isaac Asimov

First published in Great Britain 1974 by
Abelard-Schuman Limited as the
second part of *Today and Tomorrow And . . .*

Coronet Edition 1977

Printed and bound in Great Britain for
Hodder and Stoughton Paperbacks, a
division of Hodder and Stoughton Ltd,
Mill Road, Dunton Green, Sevenoaks, Kent
(Editorial Office: 47 Bedford Square, London, WC1 3DP)
by Hazell Watson & Viney Ltd,
Aylesbury, Bucks

ISBN 0 340 21785 5

Grateful acknowledgment is made to the following for permission to reprint the articles included in this book:

"Prediction as a Side Effect," July 1972 issue of the *Boston Review of Arts*. Reprinted by permission.

"The Scientists' Responsibility," reprinted from *Chemical and Engineering News*, Vol. 49, April 19, 1971, pp. 3–7. Copyright © 1971 by the American Chemical Society. Reprinted by permission of the copyright owner.

"No Space For Women?", March 1971 issue of *The Ladies' Home Journal*. Copyright © 1971 by The Ladies' Home Journal. Reprinted by permission of the copyright owner.

"Future Fun," copyright © 1967 by Local One, A.L.A. Reprinted by permision from *Lithopinion* No.6, The Graphic Arts and Public Affairs Journal of Local One, Amalgamated Lithographers of America (New York).

"Freedom at Last," as "You've Come a Long Way, Baby . . . ," March 14, 1970, and "The Age of the Computer,' as "Who's Afraid of Computers?," January 20, 1968, issues of *Newsday*. Copyright © 1970 (1968), Newsday, Inc. Reprinted by permission of *Newsday*.

"After Apollo, What?," as "After Apollo, a Colony on the Moon," copyright © 1967 by Isaac Asimov, "The Lunar Landing," as "The Moon Could Answer the Riddle of Life," copyright © 1969 by Isaac Asimov, "The Secret of the Squid," as "Spacecraft, Like Squid, Maneuver by 'Squirts,'" copyright © 1969 by Isaac Asimov. All three articles first appeared in the New York *Times*.

"The End," January 1971 issue of *Penthouse Magazine*. Copyright © 1971 by Penthouse International Ltd. Reprinted by permission of *Penthouse Magazine*.

"Personal Flight," as "Personal Flight 2000 A.D.," May 1971 issue of *Private Pilot*. Copyright © 1971 by Slawson Publishing, Inc. Reprinted by permission.

"The Fourth Revolution," first appeared in the October 24, 1970, issue of the *Saturday Review*. Copyright © 1970 by Saturday Review, Inc. Reprinted by permission of the publisher.

"The End, Unless . . . ," as "Can Man Survive the Year 2000?," January 1971 issue of *True*. Copyright © 1971 by Fawcett Publications, Inc. Reprinted by permission.

"The Perfect Machine," October 1968 issue of *Science Journal*.

TO *Michele Tempesta*
the small package
good things come in

Contents

Introduction

The fifteen essays in this book represent my various attempts to look into one facet or another of the future.

What was it that got me my start as a futurist? What made me think I had any ability to do something as esoteric as look into the mists that lie ahead and try to make out the shapes that cast those dim shadows? What made *others* think I had the ability?

It all comes down to a single phrase: Science Fiction.

I began to read science fiction when I was nine years old. I began to write science fiction not very long after and I began to sell science fiction when I was eighteen. Eventually, I even became a famous science fiction writer.

Science fiction writing, if it is anything at all, is the invention of dramatic futures. From that to being a futurist is a small step, and I took it.

When I started, being a science fiction writer lent no one any cachet of respectability. Rather the reverse. But the atomic bomb did explode, and computers arrived with their distant whiff of robots ahead, and men did reach the Moon.

It came about that there were indeed people who lived in a never-never world divorced from reality—but they were not the science fiction writers. They were all those other practical, hard-nosed, down-to-earth people who somehow assumed that things would not change; that dreams did not come true, that looming dangers always went away; *they* were the blind.

So now when we science fiction writers look into the future, people are interested in what we see.

But there is no *future*, no single state toward which we inevitably head, no place in time ahead of us as definite and unchangeable as the present and past. There are *alternate* futures, an indefinite number of them, and humanity chooses each moment which future to create in the next moment. Each choice, each decision, decides some aspect of the future and no prediction of the place towards which we are heading is any better than the prediction of the route we are going to take.

We might *assume* a route and see where it will take us. If we assume one route on one occasion and another route on another occasion, then different futures will be portrayed. Self-contradictory futures, naturally.

There might be an exciting future as in "After Apollo, What?" or a trivial future as in "Future Fun", or a dreadful future as in "The End".

Of course, some routes are more likely to be taken than others are.

If we translate the situation into an ordinary tramp across the face of the Earth any one of us would be much more likely to take a route that will lead us along established paths and through inhabited places, rather than one that will take us through rugged wildernesses. If the pleasant route leads to death and the wilderness to safety we would be right in predicting death as the probable future.

Unfortunately, as we look back through history we see that, indeed, humanity has tended to take the easy path time and time again and to suffer dire happenings in consequence.

Suppose we do it again; suppose that now, in the waning decades of the 20th Century, we take the easy path of drift. Suppose humanity, generally, allows itself to continue to breed babies in the purely biological manner it always has. Suppose it continues to allow national and regional bickering to stand in the way of global solution and to allow short-term triumph over the immediate human enemy to take precedence over, and to make impossible, the long-term triumph over the threat of the general destruction of civilization.

If we take the route of drift, then the future *is* determined and it is death. And if someone foresees that, is he a "doom-crier"? Of course! What else should one cry but "Doom!" if doom is approaching.

If, on the other hand, we take the difficult route of subduing family selfishness in favour of a strictly controlled birth-rate, and of subduing tribal, regional, and national pride in favor of a working world government, then the future will be different. It may even be glorious.

Here are a handful of futures for you, then, of different kinds. See for yourself which you might like, and which you might not like, and which you would prefer to head towards—and how.

ISAAC ASIMOV

PUBLISHER'S NOTE

The term 'billion' throughout this book is American, i.e. one thousand million.

1 · In Space

THE LUNAR LANDING

The most wonderful thing about the Moon is that it is there.

It is a large and handy world, over two thousand miles across, and only a quarter of a million miles away. It hangs in the sky, large, gleaming, and incredibly inviting, so that men have dreamed of going there long before our own planet was thoroughly explored.

Suppose the Moon didn't exist—

After all, it might not have existed. None of our immediate neighbors among the planets has anything like it. Mercury and Venus have no satelites at all as far as we can tell. Mars has two satellites which, however, are so contemptibly small (less than a dozen miles across, each) that they can be dismissed.

Without our Moon, the objects visible in Earth's night sky would all be nothing but points of light. There would be nothing to stimulate the pre-telescopic imagination into dreaming of the existence of other worlds. Points of light would scarcely have the effect on us that a visible disc would; especially a disc like the Moon, with splotches and shadows apparent to the unaided eye.

And even after the invention of the telescope, when some of the points of light would be revealed to be sizable globes, the nearest would still never approach closer than 25 million miles, fully a hundred times the distance of our Moon.

With the first step made more difficult by a hundred times, would we have the courage to dream of venturing into space?

SOURCE: "The Lunar Landing" appeared in *The New York Times Magazine* as "The Moon Could Answer the Riddle of Life", July 13, 1969.

Fortunately the Moon *is* there, a super-convenient stepping-stone into space—and the queerest part of it is that it would seem to have no business being there. There is something about it that is radically different from all other satellites in the solar system.

There are 32 known satellites in the solar system, distributed among 6 of the known planets (Earth 1, Mars 2, Jupiter 12, Saturn 10, Uranus 5, and Neptune 2) and of these 25 are small worlds ranging from a few miles to a few hundred miles across.

That leaves 7 satellites that are sizable. Jupiter has 4 of them: Io, Europa, Ganymede, and Callisto. Then Saturn and Neptune have one apiece: Titan and Triton, respectively. And, of course, Earth has the Moon.

In terms of sheer size, the Moon is next to last among these large satellites. Only Europa is smaller.

What, then, makes the Moon so unusual?

It's not size alone that counts, but the size of the satellite in comparison to the planet it circles. Jupiter's largest satellite is Ganymede, with a diameter of about 3,200 miles. Jupiter itself, however, has a diameter of 86,000 miles. The diameter of Ganymede is only about 3.7 per cent that of Jupiter.

Compared to giant Jupiter, Ganymede and all the other Jovian satellites, large and small, are only scraps. We can imagine a huge cloud of dust and gas swirling about in primordial times and slowly condensing to form Jupiter. Tiny sub-swirls on the outskirts would form the satellites. Even those swirls which produced quite sizable satellites on the earthly scale would be tiny compared to Jupiter—which has eleven times Earth's diameter.

The same is true for the other planets. Saturn's large satellite, Titan, is only 4.3 per cent of the diameter of Saturn itself. Neptune's large satellite, Triton, does a little better—8.5 per cent.

But now compare this with the Moon. Its diameter of 2,160 miles (considerably less than that of Ganymede, Titan, or Triton) is, nevertheless, fully 27.5 per cent the diameter of the Earth. Earth has a satellite that is over a quarter as wide as itself. No other planet can make such a claim or anything like it. The Moon is so large compared to Earth that, seen from a distance, the two might almost be said to make up a double planet.

Why should the Moon's size be so lopsided? When the swirling cloud of dust and gas condensed to form the Earth, why should so large a proportion of it have formed a comparatively large satellite on the outskirts, when nothing like that happened in the case of any other planet?

It's easy to say, "Well, it just did, that's all", but there may be a reason and astronomers would love to know it. So far, they haven't the vaguest notion of any explanation.

Perhaps the Moon formed in a different manner and did not take shape by way of a sub-swirl of dust and gas.

Here is one alternate explanation—

As the Moon circles the Earth, it drags the ocean water with it, giving rise to the tides. In the shallow parts of the ocean, the friction of the moving water against the sea bottom acts as a brake on the Earth's rotation. This means that the length of the day is very slowly increasing. It can also be shown from the principles of physics that it means the Moon is very slowly increasing its distance from Earth.

Hundreds of millions of years ago, Earth must therefore have been rotating much more quickly and the Moon must have been much closer to it. A couple of billion years ago, the Earth must have been rotating very rapidly and the Moon must have been practically touching it.

Can it be that when the Earth was first formed, it lacked a satellite, as is true of Venus? Can it be that as it spun rapidly, a piece of it (now the Moon) broke away? Can the Moon be as large as it is because it did not form from the cloud of dust and gas as the Earth itself did, but because it was born by a different route, breaking away from an already formed Earth.

There are serious difficulties with this theory, however.

Think of the spinning Earth. As the planet turns, the Earth's surface makes only a small circle at places near the poles; larger circles farther from the pole. It is at the equator that the surface makes the largest sweep, a full 25,000 miles, and at the equator that the surface is moving most quickly. (The surface moves at 1,040 miles an hour at the equator but only 650 miles an hour at the latitude of New York.)

If we imagine the Earth spinning so rapidly as to begin throwing off air, water, and chunks of solid land into space, it would be from the equatorial regions these would come, for those regions would be moving fastest. The thrown-off material, as it formed the Moon and traveled farther from the Earth, would continue to circle the Earth in line with the Equator. It would still be doing that today.

This would also be true if the Moon formed on the outskirts of the whirling cloud of dust and gas that was forming the Earth. As that cloud whirled, movement would be fastest in the equatorial regions and it would be there that the Moon material would cluster.

We see examples of this in the solar system. Jupiter's four large satellites all circle their planet directly above the line of its equator. This is also the case of most of the satellites of the other planets. As for the planets themselves, almost all revolve about the Sun more or less in the line of the Sun's equator, for they apparently formed at the equatorial outskirts of the huge whirling cloud that condensed to form the Sun.

The Moon is a notable exception. It does not move about the Earth in the line of Earth's equator. It revolves about the Earth (and, along with the Earth, about the Sun) more nearly in the line of the Sun's equator. In this respect, the Moon acts like a planet in its own right, and not like a satellite.

Can it be that the Moon *was* an independent planet originally? One which somehow wandered into the vicinity of the Earth and was captured by Earth's gravitational pull?

If so, where could the Moon have come from?

One interesting suggestion as to the place of the Moon's origin involves the asteroid belt. Between the orbits of Mars and Jupiter are many thousands of tiny bodies (asteroids), of which the largest is less than 500 miles across. It is often suggested that these asteroids are the remains of an exploded planet. However, even if all the known asteroids are lumped together, they make up a very small planet indeed.

Can part of the original planet between Mars and Jupiter be missing? Can it have been driven by the explosion closer to the Sun? Can it, in fact, be the Moon?

If it is, we might next ask when the capture was effected. There are geologists who argue that some billion years ago or less there are signs, in the rocks, of a large catastrophe that could easily have been the result of huge tides that swept the continents clean. These might have resulted by the sudden capture of a then nearby Moon.

Can it be, then, that the Moon was captured by Earth less than a billion years ago? In that case, through most of Earth's 5 billion year history, it existed as a lonely world, like Venus, and it is only recently that the Moon joined it.

Of course, there are catches to this suggestion, too. When astronomers try to work out the actual mechanics that would be involved in driving the Moon from the asteroid belt to the vicinity of the Earth and then having it captured by the Earth, matters become entirely too complicated for easy credibility.

In short, there is no really plausible way of explaining why the Moon exists where it does and why it should be the size it is.

But it *is* there and it *is* extraordinarily large and the lunar landing may supply us with an answer to that immense puzzle. And with other answers, too—

There are many ways in which the Moon can be useful to us and there are many ways in which lunar landings can ultimately be expected to yield us a profit. The key word, however, is "ultimately." In most directions there will be an unavoidable wait, perhaps a long one, before those profits can be realized.

It is easy, for instance, to argue that the Moon offers an ideal spot for an astronomic observatory. Without an atmosphere, seeing will be unrivaled; the Sun can be studied through all its range of radiation, and there will be an unparalleled opportunity to study the Earth itself as a whole.

Yet certainly the initial landing on the Moon won't succeed in establishing an observatory then and there. It will undoubtedly require many landings on the Moon and much in the way of tremendously complex preparations to set up an observatory.

The same can be said for almost any other practical aspect of lunar exploration. The possibilities are wonderful, but *when*—

Well, the vacuum that covers every part of the lunar surface is more free of gas than any vacuum that can be produced on Earth, and there are millions of square miles of it. Materials could be manufactured with unprecedented purity, for they would lack the gas film that is almost universal on Earth. Without the gas film they would have different properties from those which we are used to and the techniques of welding would be utterly changed, while metallic films could be layered onto other substances with unprecedented thinness.

At the low temperatures of the two-week lunar night, it would be much easier to refrigerate objects the rest of the way down to near absolute zero, than it would be on the much warmer Earth. Devices such as computers and large magnets, which make use of unusual properties that only exist near absolute zero, could be more easily constructed and more easily studied.

Complicated chemicals that are difficult or impossible to handle (such as many of those found in living tissue) might be put together in mass quantities at the Moon's low temperatures and then purified by distillation in its excellent vacuum. Thanks to that vacuum, the distillation could take place at temperatures so low that the fragile molecules would not be damaged.

The energetic radiation of the Sun in the far ultraviolet and beyond is stopped by Earth's atmosphere, but on the Moon, it reaches the surface. Such radiation could be used to initiate novel chemical reactions. The effect of radiation on the cause or cure of cancer, on mutation, on cell damage, could be studied.

Solar batteries can be used more profitably on the Moon than on Earth, thanks to the Moon's blaze of a two-week cloudless day. Research in the construction and use of all sorts of devices involving solar power can be greatly advanced, to Earth's own ultimate profit.

But when, when, when can elaborate factories and laboratories be built to take advantage of all these opportunities?

An independent colony on the Moon might be of great use to Earthmen psychologically, offering us the stimulating vision of a renewed frontier and the profitable example of a closed society making rational use of its limited resources.

But when can such an independent colony be established?

It is only natural, then, to dismiss these long-range possibilities rather impatiently and to ask whether the early lunar landings, even the very first perhaps, can be of use to us in any way. After all, we would surely like to have the first landing on the Moon something more than an extremely expensive exercise in one-upmanship over the Soviet Union.

Well, it is quite conceivable that even the first landing can earn its keep and that it can gain us enormous information.

The first astronauts to land on the Moon plan to gather up samples of the Moon's crust and bring them back to Earth for detailed study.

Those few pounds of dirt may help decide the question of the place of origin of the Moon. If the Moon were once part of the Earth or were part of the dust cloud out of which the Earth had also been formed, then the Moon's crust should be very much like the Earth's in a large variety of chemical ways.

What, though, if the Moon had originally been part of a planet in the asteroid belt, out beyond Mars?

The dust cloud out there, two or three times as distant from the Sun as we are, might well have been significantly different in chemical composition from our own much closer-to-the-Sun cloud. The planet that formed out there, taking shape under different conditions of temperature and radiation, might have assumed interesting differences in the nature of its crust.

If, then, the dirt from the Moon is markedly un-Earthlike in important respects, that would be a strong indication that the

Moon did not originate in the neighborhood of the Earth; that the Moon is an accidental and perhaps recent acquisition of our planet.

In that case, the first lunar landing would have handed us the kind of bonus we would not really have had the right to expect. It would have given us an opportunity to travel 237 thousand miles in order to explore a portion of the solar system that ought by rights to have been perhaps 237 million miles away—a thousand times as far as the Moon.

If, on the other hand, analysis of lunar material shows that in all likelihood, the Moon had always existed in the vicinity of the Earth, we need not feel disappointment. There could still be much to learn about the Moon. There are the newly discovered mascons, regions of higher-than-average density centered about the lunar seas. What and why is this? The astronauts report softer outlines on the mountains on the far side of the Moon? Could there have been more erosion on that side. If so, why?

Even though the Moon's crust, formed in Earth's vicinity, might have been very like the Earth's crust to begin with, the subsequent history of the two worlds has been radically different.

Over billions of years, the Earth's crust has been enormously affected by the action of wind, water, and living things. The result is that virtually no traces remain in being in the Earth's crust that can give us information as to what the situation was like more than, say, half a billion years ago.

On the Moon, however, the action of wind, water, and living things has been minimal. To be sure the Moon's crust has been affected by temperature differences, by solar radiation, and by meteor bombardment, but the resulting effects have been small compared to those on the Earth.

It follows, then, that the Moon's crust will be much more informative concerning the Moon's early history (and, therefore, the early history of the Earth as well) than Earth's crust is. Between what the astronauts see, photograph, and instrumentally detect, and the analysis of what they bring back, it is quite conceivable we may learn a great deal about conditions on *Earth* 2 to 4 billion years ago.

There is, in addition, very likely to be more in the Moon's crust than the minerals and crystals we expect to find in it.

If the Moon had its origin in the neighborhood of the Earth, it ought to have had its share of the common elements and molecules that go into the makeup of Earth's atmosphere and ocean. (If it was once actually part of the Earth, it might have brought some air and water with it when it broke away.)

To be sure, the smaller gravitational pull on the Moon's surface (only one sixth that on the Earth's surface) would have been ineffective in holding any ocean and atmosphere in the long run.

But what is "the long run"? For some millions of years, the Moon may have retained a shrinking supply of surface water and a gradually thinning atmosphere. Some surprising evidence in favor of that thought arises from recent studies of the Moon's surface by way of satellite photography. Meandering marks have been found on the Moon's crust that look so like dried-up riverbeds in so many details that it is hard to think of anything else they could possibly represent.

Then, too, astronauts circling the Moon have detected browns and tans among the stark black and white of the Moon's surface. Such colors usually imply the existence of iron oxide. Does this mean there was once a little free oxygen on the Moon?

If the Moon had its ocean and atmosphere once, the simple molecules in that ocean and atmosphere might have undergone changes similar to those undergone on Earth, since both worlds were exposed to the energetic radiation of the same Sun at the same distance.

The changes undergone by these simple molecules are of particular importance because they ended in the development of the enormously complex molecules of living tissue. Biochemists have been trying to duplicate the course of these changes in the laboratory, setting up small-scale analogs of what they imagine primordial conditions to have been like.

Naturally, it is difficult to tell to what extent the laboratory conditions are really like the situation on the primordial Earth and how far, therefore, laboratory results are to be trusted. On the Moon, however, we have a worldwide "experiment," perhaps, in which the changes began and then came to a halt at some midway point when the ocean and atmosphere finally disappeared.

In the samples of crust brought back from the Moon, then, biologists may find organic molecules representing a point partway toward life. Perhaps the molecules may even have reached a Moon-born life far more primitive than anything on Earth, but superlatively interesting for that very reason.

(And if the Moon had once been part of the Earth, these organic traces might have been formed from actual Earth-developed chemicals. We would, in that case, have a view of phases of early terrestrial evolution long since blotted out completely on Earth itself.)

In any case, the study of the organic molecules in the Moon's crust might advance our knowledge of the formation of life by decades when compared with the laborious advance that would have been possible by Earth-based experiments only. (Of course, one bag of rocks won't tell us all, either. We will want samples from various sections of the Moon, samples which have been parts of different environments. Our researchers will want many lunar landings—a permanent lunar base, if possible.)

Nor need we think that the question of the chemical evolution of life is something that merely exercises the theoretical speculations of ivory-tower biologists. It is quite conceivable that the ramifications of such research could be amazingly useful to humanity on an immediate medical level.

With all the biological advances made in recent decades, we still remain pitifully ignorant of the fine detail of what goes on inside cells. For instance, we know that we can alleviate the symptoms of diabetes with injections of insulin, or that we can cure scurvy by adding ascorbic acid (vitamin C) to the diet. We do not know, however, even today, exactly what it is that either insulin or vitamin C does in the body. In fact, we don't know the actual fine workings of any hormone.

That we can engage in successful hormone therapy at all is due to the fact that we imitate the action of the body. We don't know what it is doing, but if we keep exact step with it, we may get results. The fact that we don't know what it is doing, however, means that we have no way of predicting undesirable side effects until they display themselves in what is sometimes a most unwelcome fashion.

In the case of hormones, we can at least determine what the body is doing in a rough sort of way by depriving experimental animals of those hormones. What do we do when we have no such easy key? What about cancer, for instance?

Not all the research lavished on cancer in the twentieth century has told anyone exactly what goes wrong in a cancer cell. Something is wrong, obviously. The cell is not normal; it keeps growing and reproducing when it shouldn't. But *what* is wrong? What chemical reaction or reactions has or have taken the wrong turning?

We don't know, and until we find out, we may not have much of a chance of working out a satisfactory way of preventing or curing the disease.

The reason why it is so hard to find out exactly what is happening inside a cell, is that so *much* is happening. There are many thousands of chemical reactions and physical changes all taking place simultaneously and all affecting each other in the most delicate fashion. It is like trying to unravel a fine cord which has been worked into an intricate, intertwining, three-dimensional ball a foot across.

But suppose an analysis of the Moon's crust tells us what the chemistry of organic molecules is like, partway to life, or shows us the chemistry of a sub-life far more primitive than Earth's most primitive cells.

We may then get visible clues to the workings of the cell;

clues that are obscured in Earth-cells themselves by the piling up of towering complications.

Once we get the basics in this fashion, it is possible that we may get some notion as to what may have gone wrong in a cancer cell. With that knowledge in hand, biologists can then turn to Earth-cells and, knowing exactly what they are looking for, find the biochemical cause of cancer at last.

This, therefore, is the answer (or, anyway, *an* answer) to those who ask why we are spending billions on reaching the Moon, when it is so much more important to solve the cancer mystery here on Earth.

All science is one. If we push back the boundaries of darkness in any direction, the added light illuminates all places and not merely the immediate area uncovered.

It is just conceivable, in other words, that by taking the long trip to the Moon, we will be taking the shortest route to unmasking the riddle of cancer.

AFTERWORD

This article appeared in 1969, just before the first Moon landing, and a number of other such landings have taken place since. A considerable quantity of Moon rocks has been brought back on each occasion and some rocks have been brought back by unmanned Soviet rockets.

So far, so good, but unfortunately the optimistic attitude of the article has proven unwarranted, so far. The rocks have not clearly answered the problem of the Moon's early history, but have posed new puzzles.

The rocks show no evidence of water or perceptible quantities of organic matter, and the markings that seemed to be the evidence of one-time flowing water are much more likely the result of one-time flowing lava.

Undoubtedly the lack of really sensational discoveries has contributed to the loss of public interest in the Moon landings.

Nevertheless, who knows what tomorrow brings? If there is

water on the Moon at all, it would not be at the surface, which has been exposed to solar radiation for billions of years, but several feet under the surface.

So far we have only scratched a few isolated patches on the Moon's surface. I hope we continue to look.

AFTER APOLLO, WHAT?

The assumption is that we (and the Russians, too) will reach the Moon within a very few years. Even the tragic accidents that cost the lives of three Apollo astronauts and a Soviet cosmonaut do not alter that assumption. NASA has already scheduled the first Apollo flight for very early in 1968 and still looks to a Moon landing before 1970.

But it takes a few years to plan an advanced program of space exploration beyond that already existing, so we must now begin to think beyond Apollo. If we do not begin our preparations *now*, we face a halt of perhaps several years after we do reach the Moon. And with that halt might come a loss of interest in the space program generally, a loss of psychological momentum that would stretch those years of delay into decades.

The trouble, perhaps, is that the motivation behind our space program was childish to begin with. Too many Americans have aimed to "get to the Moon first," not because they wanted to get to the Moon at all, but only because they wanted to get back at the Soviets for putting up the first satellite in 1957 and humiliating us. As a result, we kept on pumping increasing billions into the program because every once in a while the Soviets would score another "first".

But the Soviets have sent up only one manned flight in two years, and that ended in disaster. We are clearly ahead of them in virtually every phase of the program (far ahead in some). That settles it, many people believe. We showed them and Sputnik is avenged. The interest in spending billions on space already is fading. We'll reach the Moon because we're committed to the

---

SOURCE: "After Apollo, What?" appeared as "After Apollo, a Colony on the Moon" in *The New York Times Magazine*, May 28, 1967.

goal (and because the Soviets might make a last-minute spurt) but after that, there are no definite plans.

Lately, there has been talk of going on to make more elaborate manned or unmanned shots to Venus, Mars, or to some comet flashing by. But is that all the Moon means to us? Is it just a relay station on the way to somewhere else, so that the space race can continue in all its expense and competitive folly?

We and the Soviets need a new motive, one that places a value on something other than scoring "firsts" and one-upping each other. Merely reaching the moon is no end in itself. Having reached it, cannot we and the Soviets and any other nation that cares to join the venture combine to explore it, develop it, and make it a home for man?

To some, of course, the question isn't "After Apollo, what?" but "After Apollo, why?"

And just as there are those who questioned the project of reaching the moon in the first place, many more will ask, "Why colonize it?"

Presumably there were also those who asked "Why?" in 1492. Once Columbus had discovered America and brought Spain great prestige, was it really necessary to go beyond that immediate gain and invest large sums of money to explore and colonize those far-distant savage wildernesses?

That old "Why?" was easily answered with material reasons: gold and silver and new and unheard-of products—maize, potatoes, tobacco, quinine—to bring back to a wondering Europe. No doubt, if the Moon had obvious resources that men valued, many would see the purpose of colonizing the Moon at once.

Yet the greatest gift of the New World to the Old was not gold or tobacco or any material resources. It was the gift of an intangible—the gift of a second chance, a new start. Later, men came to America looking for freedom as well as for precious metals and for relief from oppressive traditions as well as from poverty.

One result of that search for intangibles was the establishment of the United States of America on an equalitarian basis that was impossible for the caked society of eighteenth-century Europe to initiate, or even imitate, without vast blood and turmoil. Moreover, our size and our century-long isolation enabled us to develop a technology which the rest of the world eagerly attempts to duplicate—and which our enemies accept even more avidly than our friends.

We are now faced with a still newer New World and the opportunity arises for a segment of mankind to make still another fresh start.

To be sure, the original New World was only 3,000 miles from Europe, while the Moon is 240,000 miles from Earth. Yet such distances must be viewed in their technological contexts. The wooden hulls that pushed off from Spain more than four centuries ago faced a voyage much longer in time than the one that separates us from the Moon. The ocean voyagers were frightened men, not knowing where they were going, ill fed, poorly cared for, completely isolated from home. Our astronauts are superbly trained and cared for, highly motivated and at all times in contact with home. The advantage is all with the first men to the Moon over the first men to America.

But once the New World was reached, one might argue, men could live in it. It contained trackless wilderness, hostile natives, wild animals, but it had air and water, edible plants and animals. Men could make friends with the natives (or enslave them), hunt the animals, clear forests, and plant crops.

What can the Moon offer in comparison? It is a dead and useless world, without air and water. Its surface is exposed to the ceaseless rain of micrometeorites, ultraviolet radiation from the Sun and X-rays and cosmic rays from space. Temperatures drop to 200 below zero Fahrenheit during a two-week night and rise to 200 above zero Fahrenheit during a two-week day. Man could not live for five minutes on the Moon without artificial protection.

But what is wrong with artificial protection? The Pilgrims could not have survived their first winter in New England if they had not built some sort of shelters. On the Moon, too, it will be necessary to build shelters against the harshness of the environment. At our level of technology, it would be as easy to burrow into the Moon and carve out airtight caverns as it was for the first New World colonists to hack down trees and build houses.

The airtight caverns would provide marvelous protection. Even just a few feet of rock overhead would bar the harsh sunlight and all unfriendly radiation except for a few cosmic rays which, after all, reach us here on Earth. Micrometeorites would be warded off, and extremes of heat and cold would disappear. Underground, temperatures would remain equable at all times.

Of course, there would remain the question of water and air, indisputably a serious drawback.

At first, astronauts will have to bring their air and water with them. This might entail the prior construction of two space supply stations, one orbiting the Earth and one orbiting the Moon, like mountain camps established on the slopes of a Himalayan peak. The dependence of pioneer Moon colonists on periodic supplies from Earth is not without parallel. Three years after Jamestown was founded, the English colonists were famished and ready to abandon the project or die when they were saved only by the last-minute appearance of Lord De la Warr's flotilla with new men and supplies.

As long as Moon colonists would have to depend on an umbilical cord stretching a quarter of a million miles to Earth, the colony would be an expensive venture indeed. Yet might not the umbilical cord be dispensed with eventually? Might not the colony grow independent?

When we say the Moon has no water, we really mean it has no water lying freely on its surface. It has no oceans, lakes, or rivers, or even swamps or marshes. But it may have water just the same. Water is so common a substance all over the universe (as far as we know) that it seems reasonable to hope that the Moon will have retained at least a moderate quantity.

Moon water might consist of layers of frost in recesses of craters into which the Sun never shines. Underground, perhaps, seams of ice could be mined as we mine coal. Failing those sources, we could utilize water molecules which almost certainly are tied loosely to the mineral substances that make up the Moon's crust. The water supply would not be large enough to support a population of billions who are accustomed on Earth to consuming and polluting water as though it existed in infinite supply. It could, however, support a space colony of thousands who made shrewd and chary use of the available amount.

Such water, obtained from the Moon itself, could easily be separated by the process known as electrolysis to yield hydrogen and oxygen. The oxygen then could be used to supply the underground caverns with an atmosphere.

Minerals, too, could be locally obtained. Astronomers have long thought the Moon's crust to be essentially similar to the Earth's and everything we have learned from satellites orbiting the Moon and making soft landings confirms the belief. From the Moon's minerals, metals and ceramics could be derived, and nitrogen and carbon dioxide, too. Appropriate minerals, plus water and carbon dioxide, could be used to grow plants under artificial light. Eventually, perhaps, animals could also be fed and used as a meat supply.

Everything would have to be recycled, of course; that is, used over and over again. Men would consume oxygen, water, and food, breathe out carbon dioxide and eliminate wastes; the carbon dioxide, water, and wastes would be used by the plants to reform oxygen and food. The process is not completely efficient, but any slow leakage could be overcome by increasing the water supply from the Moon's crust and using additional minerals. Then more oxygen and carbon dioxide could be made.

This sort of tight recycling is by no means fantasy. Such systems are being developed right now since they are essential for any spaceship undertaking a journey longer than a trip to the Moon. A spaceship carrying men to Mars, for example, can scarcely count on a round trip much shorter than two years. Carrying a supply of food, water, and air—without recycling—would be quite impractical.

Recycling presents psychological difficulties. The thought of producing drinking water by evaporating urine, for instance, seems unpleasant but that is precisely what happens on Earth. At least some of the fresh water we drink was in urine once and in substances even more unpalatable. On the Moon, the cycle would be tighter and more obvious, but it would be no different in principle.

It would also demand a continuing and reliable source of energy, since it takes energy to mine the Moon, electrolyze water into oxygen, irradiate algae with light and make plants grow.

It would be impractical to carry coal or oil to the Moon as the source of energy; those materials are too bulky for their energy content. Undoubtedly an early project for any Moon colony would be to transport and assemble a nuclear fission plant like those which exist today.

But the Moon has its own source of energy, solar energy, which man on Earth is already using directly. Solar batteries that convert sunlight directly into electric currents, for instance, power a number of our satellites. Sunlight on the Moon pours down for two weeks at a time, without interference from clouds or fog. It is a tremendous and absolutely reliable energy source and one can imagine batteries eventually covering the Moon's surface by the square mile, supplying enough energy in the daylight period for immediate use and for storing and later use during the two-week night.

Then, if we learn how to control nuclear fusion, heavy hydrogen could become a far more compact energy source than solar batteries. Heavy hydrogen from Moon water might be all the colonists would need. If not, the supplementary supply could easily be brought from Earth, so compact is it as an energy source.

It seems that Moon colonists would have to depend on the perfect functioning of an intricate technology for survival. What if something goes wrong? Suppose they had to face something like the Great Northeast Blackout of 1965? Suppose a meteorite punctures a cavern or a Moonquake cracks one and the air escapes in a rush? How can men be expected to live under such imminent threats of instant death?

Men live with danger on Earth, too. They farm on the slopes of volcanoes, live on vast plains where tornadoes are common, swarm along the sea coasts where hurricanes bring floods. They mass in rabbit-warren cities where fires rage daily, and teem on highways where auto accidents kill thousands. Add it up.

Admit the possibility of technical failures on the Moon, but admit the existence of backup systems as well. Admit the puncturing of caverns, but admit also the existence of alarms, of partitioned subcaverns to limit air loss, of space suits for all personnel in emergencies. Then note the absence of blizzards, hurricanes, tornadoes, tidal waves, biting frosts, or scorching heat. Add it up.

It would not be hard to argue that, on the whole, the Moon would offer a much safer environment than the earth ever did.

But consider the unique psychological difficulties of establishing a Moon colony. Could men live out long periods, or even entire lifetimes, in crowded caverns? Could men give up the limitless freedom of the Earth, the open air, the wind and flowing water, the green below and blue above?

Not every man, perhaps, but a great many could.

Don't underestimate the adaptability of the human spirit. Men by the millions have spent their youth in some pleasant village and then hurled themselves without preparation into the maelstrom of a New York City slum. Most of them survived the vast change and a surprising number actually flourished.

The canyons of New York streets and the crowded caverns of megalithic New York office buildings closely approach likely habitats in a Moon society. Those who live in New York or any other large city are already so far from nature, so unaware of green below and blue above, so unacquainted with open air and flowing water, that the shift to the Moon will seem like a mere detail.

Yes, Moon caverns would be claustrophobic, but not every man suffers from claustrophobia. Enough men are found to crew our submarines or spend weeks in our space capsules—and enough men will be found to fill the caverns of the Moon.

As for those who eventually will be born on the Moon, they will know no other life. For them, the possibility of finding themselves suddenly on the openness of the Earth's hot-cold surface would be the true terror.

But, come to think of it, could a Moon colonist return to Earth to face that terror once he was well established on the Moon? This is a serious question for it brings up the matter of gravity—the one difference between Moon and Earth that cannot be neutralized or minimized by technology and which will require an enormous acclimation. Moreover, it seems unlikely that science can do anything at all about gravity in the foreseeable future.

The Moon is a smaller world than the Earth and its surface gravity is just one sixth the Earth's. To a man landing on the Moon for the first time, the mere act of walking will require careful adjustment, for since he will press less heavily upon the surface he will experience less friction. The Moon will seem much more slippery than the equivalent surface on Earth.

If he jumps, he will rise higher and stay in the air longer; but he mustn't be fooled, since he will come down just as hard as he did on Earth. Things will be much lighter but will contain as much inertia as always. Learning a new set of facts of life for the instant guidance of his muscles may be a matter of survival or death.

Beyond the mechanical problems of gravity, there remain unexplored physiological mysteries. How, for example, will the low gravity affect a Moonman's heart, kidneys, chemical balance? There is no easy way to answer such questions except by living through the experience. We can float a person in water, which is not quite the same thing, or subject him to zero gravity in orbiting satellites for limited periods, which is also not quite the same thing.

Though the need to endure prolonged low gravity is a gamble the first colonists must face, at least there is no reason—so far

—to expect unavoidably serious consequences. Indeed, based on what we know, there is a decent chance that the gamble would succeed.

Granted, then, that men can completely acclimate themselves to low gravity—but won't that very acclimation make a return to Earth difficult or even impossible? The sudden change from low gravity to high could be much more dangerous than that from high to low.

Perhaps the Moon colonists will adhere to a regimen of periodic exercise, designed to keep bones and muscles in preparatory trim for a return to full gravity. Or it might turn out that the colonists will see no point in returning to Earth. They may get to like the Moon and feel no urge to leave. For those born on the Moon, that would almost certainly be true.

Imagine, then, that the Moon colony is established as a going, prosperous, secure, and reasonably happy concern. What then? What's in it for us back here on Earth?

One thing is sure: the Moon will be of no use as a dumping ground for excess population. In the next fifty years, by the most optimistic estimate, we can place several thousand people on the moon. But in the next fifty years, Earth's population (if the present birth rate and death rate continue) will increase by several billion. So cross off the Moon as the solution to Earth's population explosion.

Neither is the Moon a probable source of raw materials for Earth. Nothing on the Moon is likely to be present in greater quantities than on Earth, and if an unexpected resource were discovered it would have to be valuable indeed to be worth the cost of transportation here.

Why bother with the Moon, then? Well, how about immaterial reasons?

The Moon will export knowledge. It will be a haven for astronomers, who will be able to aim their telescopes without an atmosphere to dim their view, without clouds and fog to blot it out or city lights to blur it.

Based on the side of the Moon that faces away from the

Earth, astronomers will not hear Earth's rising radio wave clamor and can probe distant depths unreachable from here. And by analyzing the Moon's ancient structure, scientists will trace the early history of the solar system and learn much about the Earth that they could not determine from Earth itself.

Is all this academic, ivory-tower research? Don't you believe it. Don't commit the fallacy of the "practical man" who thinks that because he sees no immediate use for a piece of knowledge, it has no use. All past history assures us that new knowledge, however rarefied it may seem, cannot avoid being helpful to mankind in the long run—and usually in the short run.

Nor is all Moon-knowledge and Moon-technique so very rarefied in the metaphorical sense. Some of it is clearly useful and rarefied in the literal sense.

The Moon's surface is covered with millions of square miles of a vacuum no Earth laboratory can match. Numerous technological processes can make use of so enormous and so effective a vacuum, employing devices which can easily be manufactured on the Moon but which could not possibly be duplicated on Earth.

Pure metals could be prepared without contamination by surface gases. Tiny microcircuits, built of ultrathin metallic film, could be formed with greater precision. Welding could be carried on at lower temperatures. The possibility of large-scale molecular distillation in the vacuum could permit chemical syntheses which are impractical on Earth.

The deep cold of the Moon's surface at night might make research in cryogenics (the behavior of matter at extremely low temperatures) much simpler than on Earth. Techniques for freezing living tissue could be developed, and low-temperature surgery might open new medical horizons.

An entire technology might arise on the Moon that could show us its heels. We would be reduced to stumbling after it, borrowing what we could.

And what about the space effort itself? We talk of sending men to Mars but can men endure the tight quarters of a space cabin for a one-way journey of millions of miles that will last for many months? Is it not possible that Earthmen can only reach out so far—withstand the depths of space only so long—and then break down? The change from open world to enclosed capsule might be too great to endure.

But Moon colonists would already have gone part of the way toward outermost space. They will already be in a capsule, within a giant enclosed spaceship called the Moon. They will already be living in a tightly cycled society, and under certain mitigating circumstances. Earth would still be there in the Moon's sky. Radio contact would still exist. There would even be the chance of returning to Earth if absolutely necessary. Life would not be quite as bad on the Moon, with perhaps thousands of companions, as on a spaceship with perhaps only half a dozen.

Then, once the Moon had been established as a home, its colonists could more easily cut the last connections to Earth. Next, Moonmen could, with relative ease, shift into a slightly more isolated capsule than they were used to, and explore the solar system to its furthest edge, where Earthmen themselves could not travel.

But more than that is at stake—and still more . . .

The Moon colony will be a completely new kind of society, facing completely new kinds of problems and finding completely new kinds of answers. Men will be living close together on the Moon; they will be isolated from other species of life; they will be in thrall to a controlled environment with little room for error. They will have to meet that brutal challenge in a way that might well be infinitely illuminating to the billions who will watch the process from Earth.

Just as the greatest consequence of Columbus' discovery of America was the kind of society described in Lincoln's Gettysburg Address, so the greatest consequence of Project Apollo may eventually be the kind of society that will finally answer the problems raised by man's overwhelming development of technology.

And just as the greatest export of the New World was what

we might call the "American idea," so it may well be that the greatest export of the Moon will be the "Lunar idea," whatever that may prove to be!

The answer to "After Apollo, what?" may just possibly be "After Apollo, everything!"

AFTERWORD

This article was written in 1967, over two years before the first landing on the Moon. Although no trace of water has been found on the Moon's surface, there may be water, at least in small quantities, several feet beneath the surface. That has not yet been investigated, directly.

There has been one report of water vapor being detected by an instrument left behind by the astronauts. Perhaps it issued from some nearby crack.

If there is no water anywhere on the Moon, then a lunar colony may be unfeasible. If water is present in enough quantity to be useful, then I will stand by everything I said in this article.

NO SPACE FOR WOMEN?

In one respect, and in one only, has the United States been willing to allow a Soviet space first stand unchallenged.

Did the Soviets launch the first satellite, and the second, too? As soon as we could, we were launching them higher, better, and more often.

Did the Soviets send the first man into space, and the second, too? We were right on their heels, and sent more men into space for longer intervals. Did they take pictures of the other side of the Moon first? Did they space-walk first? Did they make an unmanned soft landing on the Moon first? Quite all right! We ended by taking better pictures, making longer space-walks, and bringing about softer and more sophisticated landings.

Then, on July 20, 1969, we capped the sundae with the beautiful cherry of the first lunar landing. The men who walked the Moon were Americans and the Soviets were nowhere in sight.

Which leaves the one exception:

On June 16, 1963, a Soviet woman, Valentina V. Tereshkova, was launched into orbit and remained there for forty-eight orbits over a period of seventy and a half hours. She performed well in her solo flight, brought back her vessel, emerged in top shape and went on to marry a male cosmonaut, Andrian G. Nikolayev, who had been in space for sixty-four orbits in August 1962. Now she has a perfectly normal baby—the first and only child, so far, to have had both parents in space.

The Soviet Union has never repeated the feat; no other Soviet woman has risen above the atmosphere; but of about a hundred cosmonauts now in training, five are female.

The American space effort was unruffled by this Soviet first; and by *only* this Soviet first. NASA never budged. No American

---

SOURCE: "No Space for Women?" appeared in *The Ladies Home Journal*, March 1971.

woman was accepted for training as an astronaut then, or during the next year, or the next, or now. Of the fifty-nine astronauts-in-fact-or-in-training in the United States, not one is a woman.

Why is this so? Are women unfit for space? Obviously not, in view of the Soviet experiment. Are American women too shy to make the effort, too ladylike in some Victorian sense, too cowardly? Surely, no one can maintain this with a straight face. American women have volunteered for space in numbers, and since women have done well in any athletic event opened to them, there is no reason to think that nowhere in this 200,000,000-individual nation are there any women who would qualify.

It is just that they are not wanted. Period.

Yet they should be. In rocket vessels, compactness of contents is essential. Every extra pound that must be lifted into orbit or to the Moon must be paid for with additional capacity of the rocket engines, additional hundredweights of fuel. For that reason, none of the astronauts are giants among men. They are of medium height and medium weight. Of two men of equal ability, the smaller is preferable for space flight.

And yet women are, on the average, distinctly smaller and lighter than men. If a woman could be found who qualified as an astronaut, the mere fact that she was a woman would probably make her more suitable than a man—at least in terms of size.

Furthermore, women are biologically sounder than men. They are more resistant to stress and are less subject to a variety of metabolic diseases, including (in particular) coronary thrombosis, strokes, and other circulatory disorders. (Valentina Tereshkova, who should know if anyone does, pointed out in an article published in the spring of 1970 that women had endured silence and isolation as well as men, and that they had adapted themselves to weightlessness more quickly.) The ultimate proof of women's biological fitness is that, when protected from the dangers inherent in childbirth, women live anywhere from three to seven years longer than men.

Are women, then, intellectually inferior to men and inherently incapable of piloting a spacecraft? Considering that women in our society have, from early childhood on, been assiduously trained to avoid displaying high intelligence and to bend submissively to captious male desires (as the surest way of "getting a man") it is not surprising that most men think women are intellectually inferior and that most women accept that judgment.

But not all women do. Despite everything that custom and pressure can do, there are a number of women who are as intelligent as any man, admit and maintain the fact, and prove it by making their way through a hostile male world to achieve positions of responsibility and success. Can it be believed, then, that no woman can qualify as an astronaut? Is there no use even in making the search for one?

Or is it that they are emotionally inferior? It is sometimes asked with respect to the possibility of women serving in risk-laden positions such as that of jet plane pilots, whether anyone would be willing to chance a hundred lives or more on the judgment of a person subject to the erratic effects on her emotions of inevitable hormonal changes.

The implication is that all women become ill-adjusted neurotics now and then in connection with the menstrual cycle. Perhaps some do. Who dares say all do?

And how well would men stand up under close examination for possible emotional changes from time to time? How do men react when they've had a drink or two—not enough to be visibly affected? How do they react when they haven't had a drink but desperately wish they had one? How do men react when they've just had sex and feel somnolent? Or when they badly desire sex and feel frustrated?

If women might conceivably have their touchy moments, so do men, and to use the fact against women only is merely to disguise male chauvinism with a specious veneer. Indeed, one might argue that women's hormonal variations are usually fairly regular, so that each individual can allow for them (or even repress them altogether by appropriate medication), whereas men vary much more unpredictably, with masculine mystique often refusing to permit them to acknowledge the variation.

Yet though one could argue endlessly, the brutal fact remains that there seems to be no room for women astronauts in the American space program, either now or in the foreseeable future. If we are to talk about American ladies in space, then, we must be satisfied with a mixture of 100 per cent fancy and 0 per cent fact.

But the fancy is worth it, for there is only one way in which even the most hidebound functionary in the space agency can avoid permitting women into the rocket ships, and that is to freeze the space program at its present level or dismantle it altogether. Barring that, space exploration cannot remain an exclusively male preserve for much longer.

It is all very well to send middle-aged married men into space when they need spend only a week or two out there before returning home. For the short period involved, they can remain isolated, and leave their demure wives at home to be photographed as they sit there with hands folded, waiting for their men to return.

But what about the trip to Mars?

Do we really plan to send men to Mars? If we do, there are, at the moment, no short cuts. The trip out there will take nine months; the trip back will take nine months; and there will be a period of unavoidable waiting on the planet for the proper time to make that return. Our astronauts, in heading out for Mars, will have to reconcile themselves to being away from home for some two years.

Do we ask one man to go into solitary confinement for two years; a kind of solitary confinement that not even Coleridge's "Ancient Mariner" knew? Do we send two men to be company for each other? Do we send three? Do we send how many?

Never mind how many for a moment. Do we send only men?

Exploration is traditionally a male venture. There were no

women on Columbus' ships—but his crew spent only two months at sea, and when they landed there were women available. And don't think Columbus' men didn't know what to do with those women, even though their costumes, customs, and complexions were most un-Spanish.

This was true of all the great explorations in the past. No vessel stayed very long at sea at one time and there was always the promise of relief when land was reached. In one well-known case, that of the British vessel the *Bounty*, so successful was the relief that the crew mutinied when they put out to sea again and returned to the women.

Our astronauts going to Mars will be making a voyage longer, more isolated, and far more fearsome than any made by man before. And there will be no women waiting on Mars to console them; only the prospect of a hostile environment dozens of millions of miles from home, with the further prospect of a homeward voyage no shorter than the outer one. (Nor is this to be compared with long tours of duty in Antarctica or anywhere else on Earth, where there is at least the psychological advantage of knowing that any emergency can lead to a quick return to civilization.)

Certainly there will be men to face the Martian voyage even so, but are we to be so thin-lipped in our attitude as to fail to recognize that sending astronauts to Mars in bisexual couples might lessen a totally unnecessary part of the strain.

Are there objections? What happens if a woman gets pregnant? Let's not be naïve. These days no woman need get pregnant if she doesn't want to. What happens if there is a lovers' quarrel and efficiency suffers? Nothing worse than if two male astronauts quarrel, we can be sure.

Will the possibility of sex on board ship offend the moral bulk of America's population? Is there the thought that a crew made up of three men and three women (for instance) might engage in "orgies"? Would anyone care to think what might very conceivably happen if six men (and no women) were cooped up for two years? Would that be better?

It seems very likely, then, that if we are to have manned exploration of Mars at all, men and women will both have to be included. And if so, ought not women be trained for the task of astronaut? And why not begin now so that we can work the bugs out of the system, learn what special conditions need be evolved (if any) in the training of women, and what special problems, if any, will arise?

It is possible, though, that our sexual hangups, or our untouchable assumptions of male superiority, may bar the thought of lady astronauts at any price. In that case, we may decide to cross Mars off the list, with a grunt of grumpy masculinity. Perhaps we can argue ourselves into believing that unmanned probes to Mars and beyond will give us all the information we will need.

In that case, for the manned exploration of space, we will have to concentrate on the Moon, for that is the only object which will require rocket trips that will take weeks, rather than months or years. Only the Moon can be reached and explored in celibate austerity without undue strain.

But that depends. Might we not want to colonize the Moon? It could be very convenient to have individuals assigned to the Moon for long periods; or actually to establish an area upon the Moon that is engineered for human occupancy and made self-sufficient.

Quite apart from the chances of using the Moon as an astronomical and geological observatory, and as a physical and chemical laboratory, a lunar colony could be an unprecedented and uniquely important sociological experiment. For the first time in history, a human society would be organized within a totally engineered environment, with a totally managed ecology, and with a totally organized turnover of food, air, and water.

If such a colony worked at all, it would afford a sample of what we are coming to on Earth, where, on a much larger scale, we must learn to administer the relationship of man to the other living members of the ecology and to the inanimate environment that supports it. Failure to do so will mean utter disaster for mankind and, perhaps, for all the higher life-forms. We might (let us hope) muddle through in reasonable safety over the next half-century, and then the example set us by the lunar colony might guide our efforts to improve the muddle.

The colony would be of maximum use to us, of course, if it were a true society, with men, women, and children living on the Moon more or less permanently from generation to generation. This means that emigration to the Moon must include women. The situation will not be like the all-male crew on Columbus' ships, but rather like the families of the American nineteenth century, trekking westward in their covered wagons.

And if the Moon is to have its colony, when are we to begin training women for space flight? Why not now?

But perhaps even this is too much for NASA's sturdy masculinity. They might abandon the possibility of a colony on the Moon—at least of the kind that will feature family life. Instead, there will continue to be all-male trips, but, as technology improves and experience allows, there will be longer and longer stays on our satellite. Eventually, a party of men will remain for months, perhaps, until the next ship takes them off and replaces them.

While on the Moon, these men may not form a true colony, but they can set up an astronomical observatory, or conduct chemical experiments, or explore the features of our satellite in detail. NASA may feel, with great satisfaction, that there is much useful work to be done even without a true colony.

But then we must ask ourselves: Are there no women scientists? Are women (members of Earth's largest and most consistently discriminated against minority) even now to be forbidden a share in the great adventure? Is the accident of sex alone to deny a human being, otherwise qualified, from contributing to man's outward thrust from home?

Indeed, remember that women *do* make scientists, and many more would if the social climate were more favorable. There is no intention of arguing that woman's role in space is entirely that of sexual partner and surrogate mother, serving the men who do the real work. Under other conditions, it might well have been women—smaller, more dexterous, more resistant to stress—who would be the primary workers in space, and then it would be right to argue that men should also be sent for the comfort to be derived from their company by the women.

Neither sex alone; both sexes together! It is only because in actual reality it is men astronauts who are doing the work that the argument must stress women as the solace-bringers.

And returning to that view, if men are to remain on the Moon for months at a time, for whatever purpose, are they to be forced to be celibate? Or homosexual? It may well be that the lunar explorers may *choose* to be celibate, that they may be too fascinated by their task to worry about sex. The question is, though: Are we to make it an absolute requirement?

If not, then, when do we plan to begin training women for tasks in space? Why not now?

Retreat again! Suppose we see to it that men never stay on the Moon for extended periods. Let them go there, when necessary, stay for a short period, and return—a matter of a couple of weeks at most.

But is that all? If the space effort is to be a repeat, in endless installments, of the first Apollo ventures, the whole thing cannot survive. The lack of excitement and glamour will pall right where it will hurt the most—in the public pocketbook.

In that case, a still further retreat might be necessary.

After all (we might argue) who needs the Moon except for a very occasional venture to study its actual crust? For everything else, a large space station might be just as good; in many respects, even better.

The space around a station that is hovering several thousand miles above Earth's surface would be as airless as the surface of the Moon. An astronomical observatory would see outward just as freely and as easily. The Sun could be studied in as great detail, the planetary surfaces viewed with as great clarity, the stars studied with as great precision.

Experiments taking advantage of the short-wave radiation of the Sun, or of the hard vacuum of space, or of the cold temperatures attained in an airless space shielded from the Sun, could be performed on the space station with something of the same ease that they might be performed on the Moon.

And think of the advantages. The Earth and its atmosphere could be studied in far greater detail from a space station than from the Moon. The space station could be reached far more safely and economically than the Moon could; a virtual ferry service could see to it that men were replaced whenever necessary, reducing the strain of isolation and, therefore, the pressure to have women present.

With less money spent (after the initial investment) there would be less likelihood of public ire at the lack of periodic "spectaculars."

Most of all, a space station would afford the chance to indulge in a variety of experiments involving the gravitational field. For hundreds of millions of years, land life on Earth has flourished under the constant pull of the planet's gravity, which has never varied significantly. What would happen if gravitational fields were altered in intensity?

The field can be altered very slightly by going to the top of a mountain, but the changes due to alterations in temperature and air pressure would drown out any gravitational effect. The field can be reduced to almost zero by total immersion in water, but there would be side effects because water is much more viscous than air and restricts movement. The field can be reduced to zero in a space vessel in orbit, but in ordinary vessels the astronaut is virtually immobilized and is under unnatural constraint.

On the surface of the Moon, an astronaut would be free to move at will, but he would be hampered by a space suit. If he were in a large domed area within which he could move freely without such a suit, he would be subject to a gravitational field only one sixth the intensity of that of the Earth, it is true, but it, too, would be of unvarying strength.

On a space station set in rotation, there would be a centrifugal effect that would mimic the gravitational field in many important respects, and that would vary in intensity with one's position within the station. Men (or other life-forms), moving and exercising freely, could remain at zero gravity at the hub of the station or at increasing intensities of gravity to whatever maximum would be experienced at the outer rim of the station.

Gravitational experiments on living creatures could, conceivably, yield a great deal of valuable information on the nature of life, information that might not easily be available by any other means. It might even have direct medical benefits. (Could a man with a weak heart survive more comfortably in low-gravity fields? Could certain delicate operations be more safely performed there?)

But surely we cannot reasonably eliminate the study of low-gravity effects on women, who in some important physiological ways are interestingly different from men.

What would be the effect of low gravity on the physiologic mechanics of conception? On the development of an embryo? On the birth of a baby? On the maturation of an infant?

Is there no knowledge we can expect to gain from such studies? Can we safely trust the knowledge we gain from rats, dogs, or even chimpanzees to apply to human beings?

And if we want to study women as well as men in low-intensity gravitational fields, when are we going to begin training women for tasks in space? And if some day, why not now?

Or do we retreat yet again and, even on a space station, limit our experiments, limit the knowledge gained, limit our horizons to whatever is necessary to exclude women.

And if we do that, then we are really down to a contemptible minimum. With all the vastness of the Universe beckoning us, we will have throttled ourselves down to the region just beyond the atmosphere and to a march toward knowledge at half-speed or less; all in order to preserve space as a masculine domain.

If so, it isn't worth the money, the effort, the dreams.

Call it off; call it *all* off—or open it to the human race; the *whole* human race.

FUTURE FUN

There's the story (and if you've heard it, you can't stop me) of the Brooklyn traveler who returned from a glamour-filled trip to Paris, and who gathered his cronies about him in order to tell them of his experiences.

He gave them, in particular, the story of what had happened to him in the city's fanciest house of joy. While his friends listened, popeyed, our traveler went through every second of preliminary maneuver in careful and loving detail. Then, at the crucial moment, he stopped.

"Go on," cried out his audience. "What came after that?"

The traveler shrugged. "After that," he said, "it was the same as in Brooklyn."

—And so it will be with recreation in the future to a large extent. Some of it will be very much as it is in Brooklyn today or, for that matter, as it was in Babylon three thousand years ago.

Fun is where you find it and it is always to be found in feasting and laughing and loving and roughhousing and gambling and hiking and noisemaking and yelling and moving chessmen and chasing rubber balls and sleeping in the sun and dancing and swimming and watching entertainers and risking one's neck for foolish reasons. There are even some fortunates who find their fun in their work—as does your humble servant.

If we are to look into the future, then, and try to see what kind of recreation we are likely to have, let us agree to eliminate from consideration the kinds of recreation we already have. Much of this will continue unchanged and if some of it is to undergo modification, the alterations will not be essential. So television might go three-dimensional or movies might be piped

---

SOURCE: "Future Fun" appeared in *Lithopinion 6.* Copyright © 1967 by Local One, Amalgamated Lithographers of America.

directly into the home or a new dance may be invented—such things are trivial.

Let us instead ask ourselves, what recreations may become common in the future that are barely possible today or perhaps completely impossible? What entirely new sources of delight may we expect to be bestowed upon us (or upon our descendants) by the inexorable advance of science and technology.

To begin with, man's environment is on the point of being greatly broadened, and with that will come an accompanying expansion of recreational potential.

This has happened once before. Man was a tropical animal originally, confined to such areas as central Africa and to Indonesia (where the great apes of today are still confined); but then he discovered fire and all the realm of winter was open to him. Not only did the colder regions offer new lands, new foods, and new dangers, but (eventually) a new world of fun, too, from snowball fights to skating and skiing.

The consequences of this long-past conquest of winter have about run their course. All the world now belongs to man and settlements can be established with reasonable comfort even in Greenland and Antarctica. For now, those polar establishments are intended for scientists and soldiers, but the tourists will eventually follow, and people now alive may yet see the establishment of the Hilton-Antarctica and the Sheraton-Greenland.

But by this opening of "all the world" to human occupancy, we mean dry land, of course.—What about the sea and, in particular, the continental shelves?

If man can solve his social problems; if he can restrain his itch to set nuclear fire to himself, or breed himself into starvation, surely he will soon step back into the sea from which he sprung and the spires of his towers will begin to shine dimly beneath the waves.

Consider the dwellings of man-in-the-sea. Inside his watertight sea buildings, or perhaps under the watertight dome that will enclose an entire settlement, he will live in air and have his usual fun. But outside the dome, there will be a world of water at his disposal. That same world is at our landlubbing disposal, but we must travel to it; we are not used to it; it is a novelty to us.

To the sea-dwellers, water will be an ever constant fact of life. Children will learn to swim as they learn to walk, and scuba diving will be as common to them as hiking is to us. The sea will fill with flippered humanity hunting barracuda and exploring the drowned bottoms.

And how friendly will the new aquanauts become with sea creatures? Will boys have their pet salmon, so to speak, who will follow them about in the water?

I suspect not. Fish are not very brainy. Yet there are seabirds and sea mammals that offer a far more hopeful prospect. There are penguins and seals and, in particular, dolphins, which are intelligent and friendly.

Even landlubbing men get along with dolphins—but when men enter the sea, the friendship ought to get closer still. Dolphins are more intelligent than dogs (some suspect they may be more intelligent than men) and it may be that for the first time in history two species of intelligent creatures may meet on roughly equal terms.

And fun? Dolphin-riding ought to be a sensation that cannot possibly be duplicated on land, and I am certain that dolphins will instantly get into the spirit of the thing.

In all ordinary sea sports, even for the sea-dwellers, humans must remain air-breathing. The buildings and settlements themselves will be air-immersed and a man who ventures into the sea will do so with oxygen cylinders strapped to himself.

But will that *always* be necessary? Succesful experiments have already been conducted with water that has been oxygenated under pressure. (After all, it is not the water that drowns you, but the oxygen lack.) Enough oxygen can be forced into solution in water to support an air-breathing animal. Dogs can breathe such water and their lungs can scrabble enough oxygen out of it to support life. Dogs have remained under water for extended periods and emerged none the worse for the ordeal.

Naturally, we can't oxygenate the entire ocean, but surely we can oxygenate indoor pools under pressure. Within those pools, men can swim as water-breathing creatures and, with no equipment at all, stay immersed for hours at a time. How it would feel, I can't possibly imagine, but I suspect that it would introduce a new kind of freedom and a new sort of sensation that would be completely exciting to many.

Theoretically, we don't need a sea environment to make this form of recreation possible. (Let's call it "sub-water and breathing," or "swab," for short.) We can construct "swab" pools in Rockefeller Center if we wish and do so right now. However, persuading land-dwellers to immerse themselves in water and breathe may be most difficult. It would be far less difficult to persuade sea-dwellers—used to the friendly ocean—to do so. "Swab" may be the recreation of sea-dwellers only, however much it may be possible on land.

Then, of course, we have the Moon. Men should be standing on the Moon soon, and in one more generation, we ought to have a colony on the Moon. This, at first, will consist only of specialists, scientists, technicians, and explorers, remaining for short watches upon our satellite. Give us still another generation, however, and commercial flights to the Moon will be possible.

That might give us an answer, for the while, to the problem of "But where can one go that's really exciting?"

Just being on the Moon will certainly be fun and excitement enough for tourists. The scenery will be novel, and the sky overhead, in particular, will have the beauty of the never seen. Imagine a black sky in which there are more and brighter stars than ever we see on Earth (because there is no atmosphere on the Moon to dim them). Imagine too the Earth, as it hangs almost motionless in the sky, going through its phases like a vast and brilliant Moon.

The Earth in the Moon's sky will be four times the width of the Moon as we see it from Earth. When the Earth is full it will be seventy times as bright as our full Moon. The Earth's globe will be bluish-white and its cloud pattern will paint it in interesting spirals. Faint washes of green and brown may indicate the continents at times but I doubt that their outlines will ever be made out clearly.

There will be dangers on the Moon, of course, for without a protecting atmosphere, tiny meteorites can do harm, and so can the Sun's ultraviolet radiation. The Moon's surface can grow very hot in the sunlight and very cold during its long night. Underground, however, none of these dangers and extremes will exist and the Moon will be very comfortable.

Nor need the underground be viewed as nothing more than forever imprisoning caves. Through television receivers, views of the outside and even (properly filtered) of the Sun itself can be shown. And men will be able to emerge comfortably in the early night when the Sun is below the horizon and when the cold is not yet at its worst.

Undoubtedly, the greatest sight of all, bar none—whether seen directly or by closed-circuit television within the underground—will be those occasions when the Sun slips behind the Earth. We see such occasions from the Earth as an "eclipse of the Moon."

Once the Sun is behind the Earth, the globe of our planet will be entirely black (we will be seeing its night side) but the atmosphere all about will blaze orange-red with the slanting rays of the Sun. It will be as though we were watching a sunset scene through all Earth's atmosphere at once. And around that large bright orange circle in the sky will be the pearly streamers of the Sun's corona, visible far more brightly and clearly than ever it is on Earth. Beyond the corona, will be the hard brilliance of the stars.

Passage to the Moon will surely be at a premium in the weeks before an eclipse of the Moon is due.

But the Moon will be more than a sight-seeing paradise. It will offer active sport to Earthmen, too, thanks to its gravity. Anyone on the Moon will be pulled downward with a force only one sixth that which is experienced on the Earth. A man who weighs 180 pounds on the Earth will weigh only thirty pounds on the Moon. This will give rise to a whole new range of sensations and offer the pleasure of mastering a whole new range of skills.

Any physical activity from walking to playing football will require bodily maneuvering that will be perfected only after considerable practice. This is so, particularly, since although weight decreases, the mass of an object (the amount of matter it contains—which determines the difficulty of setting it into motion and getting it to stop) isn't changed. A medicine ball may weigh no more on the Moon than a football does on the Earth, but the medicine ball there will not at all be manipulated as easily as a football here. Its great mass will make the medicine ball just as hard to throw on the Moon as on the Earth.

Eventually, games of "Moon-ball" will have their own practitioners and their own expertise, their own rules and strategies and excitements. The "World Series on the Moon" between teams from underground stations at Tycho and Copernicus may well be followed avidly on Earth.

There will be mountain climbing on the Moon, too, less dangerous and difficult—and therefore more nearly a mass sport in potentiality—than on the Earth. This is not a paradox. The mountain slopes on the Moon are gentle and the weak gravity is easy to overcome in the upward climb. Nor do the conditions on the mountaintops grow difficult. They are airless but so are the valleys.

On the other hand, if the mountain slopes are sandy enough, the weak gravity will make them quite slippery (the smaller the force pulling you down against the surface, the less the friction). Men, using flat-bottomed canes for support and balance, may go sliding down a mountain slope for miles with all the effect of skiing, and do so (despite the necessity for space suits and oxygen cylinders) in greater safety than on Earth.

Lunar skiing may yet be the Moon's most popular sport in its early history as a human settlement.

But what about a world of no gravity at all? What about artificial space stations built in orbit about the Earth?

The purposes of such space stations will surely be purely

scientific and astronautic at first, but by the time the Moon has become a tourists' paradise (and perhaps a little "spoiled") surely some space station will be hanging in Earth's sky that will have been built primarily for recreation.

It will have to be built outside the main regions of the Van Allen belts and once placed in a nearly circular orbit out there, it will remain indefinitely circling Earth—for millions of years, if it is not struck by a sizable meteor.

Such a pleasure-satellite might have many of the ordinary pleasures of Earth and wine-women-and-song there may be essentially the same as that trio down here. (It occurs to me to wonder what novelties might be introduced into amorous techniques under conditions of little or no gravity—but let's not go into that now.)

It will also have pleasures that cannot possibly be duplicated on Earth—or even on the Moon. For instance, what about space-walking? For people who like the "wide, open spaces," what can possibly be more wide and more open than space itself? One could have a small reaction motor for maneuvering and one would have to be careful to remain in the satellite's shadow (or, preferably, to choose space-walking time when the satellite was in the Earth's shadow).

Then, for those who like it, there may be nothing quite like a few hours spent in the awful emptiness and silence of the void, when a man can really be alone with his thoughts and when he can look at the Earth's swollen body, at the Moon's more distant shape, and at the quiet stars.

You might imagine that our space walker can indulge in acrobatics, but if he does he will not be conscious of them. It will be the rest of the Universe that will seem to jump about and he himself may merely become dizzy.

For acrobatics, I suggest another recreation that can probably be found only on our space station. Why not a large empty cavernous room somewhere in the station, filled with air—possibly under pressure, to make it denser.

A man's arms can then be outfitted with "wings" for maneuvering and he can launch himself into space. The sensation of air about him will give him the feeling of movement he could not have had in empty space, and his wings will give him a personal control of his maneuvering far more delicate than would be possible by means of a reaction motor.

In short, he would be flying under his own power and, with sufficient practice, he could gain the proficiency of a bird on Earth. There would be others using the "fly-room" at the same time and a whole new spectrum of fun and games would become possible.

How about three-dimensional square dancing? Why not have two couples do-si-do-ing at right angles to each other—one couple does it right-to-left-to-right; the other, up-to-down-to-up.

Would this not be "cube dancing"?

But is nothing left for us Earth-lubbers down here? Are the new excitements to be found only in sea and space?

Not at all! The greatest new world of all lies within ourselves. There are mental recreations as well as physical ones.

Consider chess—an endlessly fascinating game which involves not the muscles but the mind. It is at present of limited interest because only a few people have the temperament and ability to make worthwhile chess players.

That can also be said about baseball, yet baseball is popular because millions who could not play except in the most amateurish fashion are willing to spend hours upon hours in watching professionals. I understand there are people in the Soviet Union and elsewhere who will similarly stand and watch large chessboards on which the moves of grand master tournaments are displayed, but this can never grow as popular a spectator sport as such games as baseball or soccer.

The trouble is that where ballgames are fast and simple, chess is slow and subtle.—But computers can play chess, too. Even as I write, a computer at Stanford University is playing another at the Institute of Experimental and Theoretical Physics in Moscow.

Computers are pretty poor chess players at present, but they will improve. Perhaps the day will come when computers will play chess at great speed and men will watch large reproductions of the swiftly changing patterns on chessboards with interest and absorption. Great games can be repeated in "slow motion" and analyzed. We could become a nation of "chess watchers."

And why just chess? New games can be invented—deliberately complicated ones with tantalizing rules that would be far too difficult to serve as efficient recreation for men, but which could tickle the fancies of computers.—Three-dimensional chess, for one thing.

To be sure, computers can't play by themselves. They have to be "programmed" by men: the rules of the game must be fed into them together with a description of desirable courses of action. Computers may start as fifth-rate players indeed, but if they are programmed to modify their play in accord with experience, they can improve just as human beings do. Computers may even become more proficient than any human being at some game in which they are designed to specialize.

We may eventually have a whole family of computer games to serve mankind.

You might ask if this is indeed the sort of thing to which one ought to apply computers and programmers, and the answer is a clear and loud, "Yes!" In the first place, what is wrong with entertaining human beings? Man must be amused as well as fed or in what way is he different from an ox?

Then, too, computer games will serve a purpose. We call them "games" but any decent game has an underlying order and pattern which, when properly studied, can serve as contributions to mathematics. To program a computer to play chess is a way of testing mathematical techniques that can then be applied to more "serious" problems. And programmers who whet their mathematical fangs on chess will find them all the sharper in other directions.

Horse racing "improves the breed," they say. Well, game programming will improve the breed of computer and programmers alike.

But even the computer is an artifact. What about man's mind itself?

Already, we are playing games with the mind (not altogether safe games) that would have been unthinkable a generation ago. The psychedelic drugs offer us a means to a new direct form of entertainment, *if* they can ever become safe enough to use, and if we can learn enough to use them with full control.

Why read of events in a book, or listen to them on radio, or watch them on television, when we can *live* them in the mind? The time may come when a man may live a complete hallucination to order; live another life far more colorful and exciting than the one of every day; live a fantasy, designed and created to order, one that is guaranteed to be as risky and dangerous as might be desired, but to end safely at last.

Many, it may well be, will prefer such dream lives to reality; dream lives in which they are much more capable and fortunate than they really are, and in which other people are much better-looking and agreeable than they really are, and in which all really happens for the best in the long run. They may prefer to remain in the fantasy more or less permanently.

Is this a horrible eventuality? I don't think so. In the automated world of the future, in which there will be, in any case, a superfluity of human hands no longer needed for daily work, why shouldn't some retire into permanent fantasy if they wish?

It may even be that the actual powers of the human mind itself will be intensified (with or without enhancement by mechanical device) so that men may finally learn to be telepaths.—Some more so than others, of course.

Who can imagine what fun it might be to think to one another rather than to talk? What wonders of the human spirit may emerge when each individual is no longer imprisoned by a wall of flesh, but can commune directly with others?

It may be, in fact, that this is the ultimate pleasure and recreation, the purpose toward which all of intelligent life has been tending since the beginning. The delight of direct communion may be such as to sink all other pleasures to nothing.

It may even be that, just as I sit here now trying to imagine the pleasures of the future, some centuries hence another man may sit and try to reconstruct, in sorrow and sympathy, the miseries of a past in which billions of human beings wandered lonely, seeking in the wildest physical and mental activities that pleasure which could only be obtained through the touch of the mental tendrils of a loved one.

AFTERWORD

When one writes as much as I do, and in as many different fields, there is bound to be interesting cross-fertilization at times. Four years after I wrote this article I wrote a science fiction novel, *The Gods Themselves* (Gollancz, 1972). In the last section, the scene was set on the Moon and if you will read it you will see that I did not forget what I had written (without any thought of fiction in mind) in "Future Fun."

2 · On Earth

PERSONAL FLIGHT

Air flight has become so common nowadays that it is a rare person who doesn't fly for even the most trifling reasons. Yet at the same time, flight is becoming less and less personal. As planes grow larger and more elaborate, the very sensation of being in the air diminishes.

One can get into a plane as large and as crowded as a movie theater, fitted out with the screens and motion pictures that go with the simile; with a dozen smiling hostesses and with luxury food and drink. The earth sinks down, moves backward, and comes up. It is the earth, somehow, that has shifted, while the movie theater one has inhabited has remained motionless.

But the very success of the luxury is enslaving. The traveler must go only at a time when hundreds of other people want to go. He must travel to a place where there is enough room and enough elaboration of facilities to make it possible for an object the size of an ocean liner to rise into the air. He must descend finally in another such place. He is far from home when he starts his flight and far from his destination when he ends it. It may well be that battling his way from starting point to airport and from airport to end point takes far more time than the flight itself.

Short flights increasingly lose their time-saving advantage.

Is there no way in which one man can go where and when he pleases? Can he not start from a point of his own choosing and land in another point of his own choosing?

SOURCE: "Personal Flight" appeared in *Private Pilot* as "Personal Flight 2000 A.D.", May 1971.

Is there no flying automobile?

In science fiction we have taken up the problem for a couple of generations now and one romantic answer was to eliminate all the trappings and frills of flight and reduce it to the very minimum required to move a man—a reaction engine strapped to his back.

The man riding the jet blast rises, travels at will, and descends. He would be the lineal descendant of wing-flapping Icarus in the Greek myth or the man in the soaring glider of the late nineteenth century.

But though the jet-propelled man is a practical possibility, he is the aeronautical equivalent of the motorcyclist, who has reduced the equipment of a multi-passenger bus to a pair of wheels and an internal-combustion engine.

Individual jet propulsion might be suitable as a sport, and valued by the young for the cachet of danger it carries with it, but it probably can never be a suitable form of mass short-distance transportation. The human body riding the individual jet, like the human body riding the individual wheels on a motorcycle, is insufficiently protected, and the opportunities for luggage, or for a fellow-passenger, are limited.

We need not the air equivalent of the motorcycle, but the air equivalent of the automobile.

Would that not be the small one-man airplane? Yes, but for the difficulties of the takeoff and landing. If enough room is to be allowed for taxi-ing into the air and then for taxi-ing to a stop, the whole thing becomes impractical in just those areas where such transportation is most required but where there is no room.

Helicopters, then? Perhaps, but they are slow and vulnerable, the equivalent of the Model T Ford. Their one advantage—the ability to take off and land vertically—will be the property of properly designed conventional planes.

Once the VTOL ("Vertical Take-off and Landing") plane is ready for mass use, we will then finally have the equivalent of the automobile in the sky. We can expect to see a world in which it will be common for the commuter to rise out of his backyard and land in a VTOL parking spot—or in a friend's backyard.

Undoubtedly both take off and landing spots will have to be specially designed to withstand the shock of departure and arrival, and not any place will do, but it may be taken for granted that any important advance in technology requires the spin-off of subsidiary improvements. The use of the automobile meant the arrival of garages, paved driveways, and service stations.

The advantages of commuting by air, rather than by land, are considerable. The air is a three-dimensional medium, without (given enough height) physical obstruction at any point. The room available in the air is therefore, to begin with, much greater than on the ground.

This means that at the start, at least, the VTOL commuter will have a great sense of freedom and will be able to make a beeline for his destination.

Yet that will represent a transitional period only, for as VTOL travel becomes more common and the airways become thicker with the flying automobiles, the crowding will become serious, three dimensions and all. There will then have to be roadways through the air, as there are along the ground.

The air is a more elusive medium than the ground and the roadways will have to be correspondingly more abstract. They will consist of radio beacons. Each commuter will have to fly his radio beacon to the point of destination and each will have to make sure it is clear first.

As time goes on, there will cease to be the necessity of the commuter's choosing and clearing his own beacon. We can look forward to the computerization of the process. The one- or two-man VTOL will become, in a sense, a flying robot. Each commuter, having dialed (or otherwise indicated) his destination, will be informed of the exact time his beacon will be free.

In advance of that time, he will get into his VTOL, adjust its controls for proper receipt of the signal, and open his book or newspaper, or just prepare to stare thoughtfully or sleepily at the scenery. At the appropriate time, the VTOL will rise into the air, make its way along the beacon, and sink to its destination.

If the commuting is at a fixed daily time, a particular VTOL may have its regular run reserved. If an emergency trip must be made, there may well have to be a wait, longer or shorter, for an appropriate beacon to become available and our friend may then indulge in the favorite occupation of commuters everywhere—that of muttering remarks concerning the inefficiency of the system.

The effect of VTOL would have enormous effects on ground travel, too. The latter would not disappear, of course, since it would still offer the most economic method for transporting goods and heavy freight. In fact, the decline in the number of automobiles on the highways and, particularly, within the cities, would make the transport of goods so much more efficient as, in itself, to pay back many-fold the investment in mass VTOL.

By draining off those who would travel personally, and sending them into the air, it would be easier to develop and expand facilities for mass transportation for those who, for any reason, wish to travel by land, so that a healthy equilibrium between land and air travel may be set up which will shift back and forth slightly in accordance with the season of the year and the condition of the weather.

Still further, the coming of the VTOL would encourage a vast decentralization of mankind.

Let us suppose, to begin with, that when the twenty-first century opens, mankind has achieved population stability and firmly recognizes the fact that Earth's population must be allowed, humanely, to decline to some optimum level. (Without such an enlightened population policy, it is impossible to look into the future and see anything but catastrophe.) Let us suppose, further, that the gathering intensifications of the problems facing us today have brought about the equivalent of world government.

The next step, then, would be somehow to untie the vast knots of mankind that have coagulated into unwieldy and decaying metropolitan centers. (These centers are already visibly breaking down and will be in even more serious case a generation from now.)

In an earlier period, the automobile had made possible a step toward decentralization. It had built up the suburbs and drawn men away from the centers of cities. This influence, useful and beneficial in itself, was overtaken by population increase generally, so that the cities grew faster than their populations could escape and the suburbs, one by one, were steadily engulfed by the problems of the metropolis.

The coming of the VTOL, *combined with a sane population policy*, gives us another chance. Men, traveling more quickly and more conveniently, can base themselves farther from their business. They can be farther from centers of amusement and culture without feeling uncomfortably isolated. The city will begin to spread out.

So, for that matter, will centers of amusement, culture, and business. When it becomes easy to travel several hundred miles, and child's play to hop several dozen, a dozen moderately sized cities, well spread out, will serve the same purpose as a single monstrous metropolitan center. In a sense, the city will disappear and nothing but suburbs will remain. Or, to put it another way, the suburbs, knit together by VTOL, will *be* the city, with all its advantages and few or none of its disadvantages.

The developing computerization will even tend to reduce the absolute need for VTOL travel for business reasons. More and more it will be information that is sent back and forth electronically, while human bodies are allowed to stay put. There will be less necessity for a dozen men to gather physically, when their images can do so by closed-circuit television (thanks to communications satellites). There will be less necessity for blue collars, white collars, and gray flannels to gather at an industrial plant when its machinery is fully automated and it can be controlled by electronic monitoring from a distance.

Does this mean that transportation will cease altogether?

Not at all. It means that *imposed* transportation will diminish, but people don't travel for business only, surely. They will still travel for pleasure, whether to see the sights, or each other, and do so in greater comfort thanks to the unclogging of the airways that the diminishing of business travel will bring about.

With information spreading the world over by radio-borne facsimile and with people borne all over by radio-guided VTOL planes, the world will be small indeed.

It will be small enough for its inhabitants to come to know each other intimately; small enough for its population to grow to know all Earth's corners and learn to love all of it and not just their own region; small enough for the land to become a "global village" and learn to live in peace; small enough for its ecology to be grasped as a whole, to be understood and cherished; small enough to form a unified base from which men can reach out and explore the solar system and the wide Universe beyond, not out of national pride and fear, but out of the calm pride of a species who, having organized their own planet, at last, in peace and harmony, seek new frontiers against which to sharpen their wits and stretch their abilities.

AFTERWORD

In my capacity as writer I am completely different from myself in my capacity as myself. In real life, for instance, I am coarse and ribald; as a writer, I am pure and refined. In real life, I can't bear the sight of blood and turn queasy at the description of any ailment or accident, however mild. As a writer, however, I can discuss the most ferocious physiological misadventures without turning a hair.

Well then, in real life, I won't go on planes out of nothing more than sheer cowardice, but as a writer, I can turn out the foregoing article with ease. So for those of you who happen to know I won't fly, don't bother to ask me how I can sound so enthusiastic about all those futuristic aeronautical devices when

I know I'll never use them.—No contradiction! It's Asimov the Writer who did the article, not Asimov the Me.

I admit I wasn't honest enough to tell the editor of *Private Pilot* that I didn't fly, when he asked me to do the article; but I'm telling *you*, am I not?

FREEDOM AT LAST

A baby girl, born today (1969), will be a young lady of twenty-one in 1990—and she may well be free in a sense we can scarcely grasp now.

How can we know that? By deciding on what will inevitably come to pass by 1990 and attempting to judge the sure consequences.

Some things we cannot know of the future and cannot, in good conscience, pretend to predict. We have no way of knowing who will be Pope in 1990 or who will be President of the United States or even who will win the World Series.

Some things *may* have happened by 1990. We *might* have had an all-out nuclear war by then; there *might* be peace in the Middle East; there *might* be a permanent colony on the Moon. We can only hope or shrug in such cases.

An all-out nuclear war may reduce mankind to scattered groups of degenerating survivors and what will happen then can scarcely be predicted now. We can only cross our fingers and work to prevent such a dreadful possibility.

Peace in the Middle East or a permanent colony on the Moon might be most interesting and desirable, on the other hand, but neither might immediately affect everyday life in the United States, even if it came to pass.

We must ask ourselves what events of great consequence will be *sure* to happen in the United States and in the world, barring unforeseen and unwanted worldwide catastrophes?

One thing, of course, comes to mind instantly—a further increase in population. The huge strain placed on society by the result of that further increase imposed upon an already over-

---

SOURCE: "Freedom at Last" appeared in *Newsday* as "You've Come a Long Way, Baby . . .", March 14, 1970.

populated world will have shocking consequences and it ought to be possible, perhaps, to see how those might affect Miss Twenty-One of 1990.

We are 3.5 billion strong on Earth today. There will be 4.7 billion of us on Earth in 1990. The population of the United States will stand at 300 million, nearly a third of a billion. There will be twice as many Americans in 1990 as there were human beings on all the Earth in the time of Julius Caesar.

We can be quite certain, then, that by 1990, the debate over the desirability of birth control will be over. The question of its morality will have become quite academic. The necessity for halting the population upswing will be denied by no important government figure, nor, for that matter, by any notable religious leader, either.

India and Indonesia, and perhaps other sections of the world as well, will have supplied the necessary object lesson. The great and grisly famines of the 1980s will be fresh in every mind and will have settled the matter.

The old biblical injunction of "Be fruitful, and multiply, and replenish the earth" will no longer be applied to conditions so radically different from the times in which the Book of Genesis was written.

Mankind *has* been fruitful, he *has* multiplied, he *has* replenished and over-replenished the Earth. He can go no farther without utter disaster. It is not just numbers alone, but what those numbers will do; the strain on the food supply and on the very fertility of a soil turned more and more desperately to a single-minded absorption in the growth of food crops; the swallowing of the resources of the world; the outpouring of pollution and poison; most of all, the vanishing of the dignity and comfort that comes with space and privacy.

Clearly, age-old assumptions and attitudes will have to change, despite all protests—particularly in those areas of the world where a large part of the population is kept intimately aware of world problems. It will be in urban America where changes will be most rapid and radical.

Large families in 1990 will seem anti-social. Motherhood will be an uncomfortable, rather than an idolized, condition, and every child will seem a threat. Unpalatable? Of course! But with the actuality of famine staring every person in the face it will be necessary to turn to the unpalatable because the alternative will be even less palatable.

The impact of all this on the young lady of 1990 is clear enough. The age-old feeling that a girl's chief role in life must be that of an eventual wife and mother will have been lifted. She may be a wife, if she chooses to be; and a mother (within careful limits) if she chooses to be. But she may also be neither, if she chooses that, and society (for the first time in history) will not sneer at her. There will be no pressure on her to breed, by way of the superior status societies almost invariably place on the mother as compared to the non-mother and on the wife as compared to the non-wife.

This freedom to be a non-wife and non-mother, if she chooses, will lift a tremendous burden from her shoulders, one that crushes down upon every woman now and goes unnoticed only because it has been there for dozens of centuries.

Once motherhood no longer comes to represent the highest and noblest status to which women can aspire, there will be a readjustment of many mother-related outlooks.

Sex will no longer be a quasi-holy sacrament because it involves the potentiality of motherhood; or a base and wicked act under those conditions where it threatens the security of motherhood. It will be neither, but will become one more variety of pleasure to be discussed freely and experimented with joyously. There must be the proper forewarning of the possible dangers of disease or pregnancy but that is not earth-shaking. The pleasures of the table are enjoyed despite the forewarning of the possible dangers of food poisoning and obesity.

Through many centuries, our culture has made every effort to restrict the sexual activity of women to single partners, at least partly because of woman's role as a potential mother. A father would naturally wish to be sure that the son who inherits his property is really his and once there is this economic reason for pre-empting the sex life of a particular woman, it becomes intertwined with masculine self-respect as well. Once the economic factor is diminished, the occasion for jealousy will diminish as well.

There will be greater variety to the sex act as well. In the world in which our traditions were formed, child mortality was always high and it took maximum sexual effort to bring enough children into the world to supply the eventual peasants, artisans, and soldiers to keep the society going.

In such a case, sex had to be channeled toward childbearing specifically. Those methods of achieving sexual satisfaction which had no chance of leading to conception were branded immoral and illegal. Masturbation, homosexuality and oral-genital contacts were, for instance, labeled unnatural and perverse not because they were really unnatural (they were all too natural, or it would not prove so impossible to stamp them out even with all the rigor of the law and all the threat of hellfire) but because they are effective birth control methods.

Once motherhood is de-emphasized, birth control becomes easier. Instead of requiring abstinence (a lost cause) or some sort of "rhythm" method (very risky), or using some sort of mechanical or chemical device (always costly, always troublesome), there would be the added method of merely encouraging what is, in any case, a natural impulse.

Practices previously considered perverse because they do not lead to conception, will become tolerable and even praiseworthy for exactly the same reason. (Naturally, those "perverse" acts which involve physiological harm, notably sadomasochistic practices, will continue to be subjected to cultural disapproval.)

The young lady of 1990, then, sexually unforced and unchanneled, will undoubtedly have been experimenting with sex for years, will know exactly what she enjoys and wants, and will have no hesitation in asking for it.

Nor need we look forward, necessarily, to a 1990 that will consist of a perpetual sexual orgy, or expect that the young lady will be exhausting herself in search of sexual novelty.

The contrary is much more likely. With the taint of the forbidden removed, a large part of the false glamour of sex will be gone. She can engage in sex when she has the opportunity and inclination but she will also have the inestimable freedom of being under no compulsion to engage in sex by making an opportunity and forcing an inclination. She doesn't have to do it merely because it is so exciting to be wicked; she doesn't have to do it just because no one's looking and now's her chance; she doesn't have to do it just because it's the fashionable thing to do. She can afford to do it only when she wants to and how she wants to.

With 1990 more sexually permissive, it will become less sex-wasteful.

It is not only population that is sure to increase through 1990 (always barring an all-out nuclear war). It is also the level of technology.

This means that the need for "unskilled labor" will continue to decline. The necessity for brute muscle will be gone, at least for purely economic purposes, though men will undoubtedly still enjoy the interplay of muscle in sports, games, and exercise.

To fulfill one's economic role, it will no longer be necessary to dig a ditch. One need merely push a button that will activate a machine to dig a ditch.

The work of the future will be largely creative, instructive, diversive, administrative. Creative individuals will include the scientists, writers, and artists, of course—women as well as men. The instructive individuals will include the teachers—women as well as men. The diversive individuals will represent the whole gamut of those who amuse an audience from ballplayers and lecturers to musicians and comedians. (In the world of 1990, show business will be on its way to becoming the largest industry in the world, for leisure time will have to be filled.) Here, too, there will be women as well as men.

What about the administrative class? Here we will have all the people who, in one way or another, guide the political and industrial machines of the world. The leaders. This is the one group which at present is largely closed to women. Will it still be closed to women in 1990?

Perhaps not. In the future, the field of administration will come more and more to mean the care and feeding of computers.

This means that in the new administrative aristocracy, for the junior executives of tomorrow, the important techniques will be those of arranging papers, flipping switches, manipulating contacts—

In all of these things, the small graceful fingers of the woman may well be nimbler, speedier, and more accurate than the larger and clumsier fingers of the man.

But will she be intellectually capable of dealing with computers, with business machinery? Will she be able to make decisions?

Let's see—

In a culture in which human muscle is the prime source of energy and power, the greater level of masculine strength places man in a naturally dominating position. The brute force of lumberjacking, the grim endurance of mining, hunting, farming, all require the strength of a man. Where battle requires the human arm to shoot an arrow, hurl a lance, wield a sword, lift a shield, only men (and strong, well-trained men at that) need apply.

Women were bound to take a subsidiary role, then, and seek for economic survival in the shadow of and under the protection of some man. The kind of behavior expected of a woman, under such conditions, is precisely that which would best attract and keep a man.

"Womanly" meant passive, submissive, and docile. A woman took orders and brought her husband his pipe, slippers, and, when required, she brought him her body. Under ideal conditions, she was not supposed to enjoy sex (that wouldn't be ladylike), since if she did, she might make inconvenient demands upon her husband.

Even more insistent is the demand that "womanly" mean unintelligent. It is essential that a man, endlessly humiliated, perhaps, in his contact with other men, find himself superior at all times to women, mentally as well as physically. In this way, he automatically remains in the top half of the human race.

Women co-operate, since they must, and through generation after generation have carefully taught themselves the ladylike art of being stupid. They have learned to be incapable of adding up a column of figures, at a loss to follow the intricacies of politics, made faint by the necessity of understanding science. Indeed, they have trained themselves to be proud of their incompetence at anything intellectual.

The reward for all this is the privilege of being thought "cute" and of having a man patronizingly reward them with such nonsense services as the removal of a hat in the elevator, the offer of an arm at the curb, the holding of a door, and the kissing of a hand. All this in exchange for abandoning the intellectual role that is the highest and proudest mark of a human being and for being condemned to a lifetime of degrading service.

The punishment for the woman who does not choose to cooperate is quite marked, and is visible even today. The "career woman" is harassed at her work by resentful men who consider her a menace to the security of their own position. They are harassed to an even further extent by other women who suspect careerists of being unwomanly and of dismissing their "proper" role as wife and mother—but who perhaps are really resentful at being made to feel the degradation of their own position.

But the young lady of 1990 will be free of condemnation to such degradation. With motherhood no longer an insistent goal, she will be encouraged to find other goals. With brains increasingly at a premium as the years of the future pass, there will be less and less readiness to dismiss half the human race. Half-speed ahead will just not be enough.

And again we can ask, Will women meet the intellectual challenge? Why not, once she is no longer brainwashed into imbecility as the price of catching a man?

Not so long ago, for instance, bank tellers were almost always men and we were all convinced that that was only reasonable since women could scarcely be expected to add up columns of figures and get correct answers. But the employment spectrum of today is such that it is hard to get men to work as tellers, women are hired, and—surprise, surprise—they can apparently handle columns of figures perfectly well.

The two freedoms—economic and sexual—will reinforce each other. The young lady of 1990 will not need to sell herself to any man because that is the only legal and respectable way of having sex and children; or because that is the only socially acceptable way of achieving economic security.

She need only give herself to a man if she wants to and only for as long as she wants to; and all she requires in return is that he give himself to her on the same terms.

This does not necessarily mean the end of the family as an institution. It does mean the end of the family as a prison.

Men and women can still form an association that may prove loving and life-long; but will not be forced to form an association that is life-long even after it ceases to be loving and therefore becomes hating.

The chances of a happy marriage will very likely increase and the number of *successful* families grow larger once the element of force is removed. Nor need a marriage begin for reasons of desperation and panic. As long as marriage is not necessary for status and as long as casual contacts are available for sexual needs and work for economic independence, our young lady need not end up with a man who is merely the best of a bad lot, or merely the only one who asked.

Nor need children (and there will be children even in 1990, for an average of two per couple will allow for the necessary slow decline in population—considering the fact that some will always die before puberty) be subjected to the insecurity of a forced and hate-filled marriage. It is generally agreed that this is worse for the child than a broken marriage and in a society where women are economically independent and where lack of a husband is no bar to either status or respectability, the man can go.

Where freedom stands in greatest danger of being lost is in the anonymity, the cipherism that may easily mark the individual in a super-crowded society. To prevent that and to retain the joy of life, every form of self-expression should and would be encouraged.

Take clothes, for instance—

There is a wide difference in the traditional clothing of men and women in most cultures today and in the past, because the two sexes play such different roles in society that it is necessary for some visible distinguishing badge to be insisted upon. You don't want to mistake a man for a woman for the same reason that you don't want to mistake a chairman of the board for a janitor.

A young woman must (in our culture particularly) advertise her womanhood because of the never ending competition to trap or hold a husband. And so we find our advertisements featuring clothes that are relatively tight, brief, and sheer, designed to show off the smooth curve of the buttocks: lift, separate, and (if necessary) pad the breasts; outline and redden the lips, and so on. It is the eye-catching exterior designed to attract the attention of men at as great a distance as possible.

But where women can find fulfillment in a role other than that of wife and mother, where a man need not be lured into a total, permanent arrangement, but allowed to enter a casual, temporary one, the advertisement need be less strident.

Again it is a matter of freedom. The young lady of 1990 may, if she chooses, dress in the acme of what we would today consider femininity, but she need not. She is not bound in the prison of sexual advertisement and may therefore use her clothing as an expression of individuality. (This will be true of men, too, of course.)

The range of variation in clothing and other adornment, both in color and styling, will be much greater than it now is from person to person, and, for a given person, from time to time. Clothing will probably be valued for its versatility and flexibility and Miss 1990 will be prouder of the creativity and ingenuity with which she can bend the dress to fit her mood than of the dress itself. (With added leisure there will, in any case, be considerable pressure to develop creativity in all ways, for the good of society as well as that of the individual. If a novel adornment is perhaps less exalted than a new symphony, it will still have its satisfying place in the scheme of things.)

The enhanced individuality may well be sexually useful, too. What Miss 1990 advertises is not herself as a generalized female, a role she may fit rather poorly on closer examination. She will instead proclaim herself as an individual. This may attract fewer moths to her flame but those who come (unconditioned to expect only certain standards of clothes and makeup) will be reaching for the genuine article and will be more worth having.

Naturally, with bizarre and individual costumes tolerated for both sexes, it will become less easy "to tell the boys from the girls." The more conservative members of our present society are already disturbed about this and yet one wonders why. Why does it disturb anyone that he cannot tell whether a stranger a block away is a boy or a girl? What does he have in mind?

Such distinctions, in a world of increasing economic and social equality of the sexes, would be of interest chiefly for immediate sexual reasons. Undoubtedly it will always be possible for a young man to recognize a young lady (or another young man if he happens to be in a homosexual mood), and for uninterested strangers the distinction is really irrelevant.

In the world of 1990, the colleges and universities should have become citadels of individuality. With more leisure than ever and with less formal work, it is more important to encourage creativity than to insist on drilling irrelevant and identical knowledge into all heads, of whatever intellectual shape they may be.

The necessary courses to fit the ordinary person for a place in a computerized and automated society may be brief and there will be (we can hope) enough who will, of their own accord and for their own delight, wish to undertake the extensive and intensive educational program that will lead to qualification as scientists and as other specialists. This will free the others for doing *their* thing.

For all, the important matter is to learn to live as an individual and to know how to express one's self in such a way as to delight the ego and, if possible, delight other egos in the neighborhood as well.

Miss 1990, in her college, will not be there in order to attend set courses, but in order to find an environment which will best encourage her to be in the learning mood—for whatever she chooses to learn. It may be knitting rather than Latin, or it may be anthropology rather than television repair—that's her business.

And the true passing grade will mark not the successful parrot, but the successfully happy person.

A thoroughly controlled birth rate then (and without one, all, *all* is lost) ought to mean freedom for the young lady of 1990; freedom beyond our present dreams; freedom not to be a wife and mother; freedom not to be a sexual chattel sold for economic security; freedom to be an individual in sex, in work, and in play; freedom, in short, to be the one unique human being one is, and no one else.

THE AGE OF THE COMPUTER

We live in a privileged age, for today, as never before, we have a choice of dooms for every taste. We can look into the future and see the planet poisoned by pollution; or rent by racial conflict; or suffocated by overpopulation; or devastated by nuclear warfare.

All these cruel visions of the future are horrible enough and, alas, possible enough, to warrant the strongest efforts to avoid them. And, for my own part, I see one possible highway to safety.

We live in the beginning of the age of the computers and, barring violent catastrophe, the age is sure to grow and intensify. There will be more computers each year, doing more tasks more intensively, and requiring ever less interference from mankind.

To some people, this seems like one more variety of doom. The computers seem to them to be cold, hard, and impersonal; they are mechanical minds without warmth, sheer mentality without sympathy, absolute justice without mercy.

But is this necessarily so? Certainly, not so far.

As yet, the computer is primarily a tool for solving mathematical equations by the rapid manipulation of electric currents. We have had tools to aid our thinking before this. A slide rule is a simple example of such a tool. Pencil and paper is a still simpler one, whenever we use them to multiply two large numbers that we cannot multiply in our head.

In fact, the very notion of mathematics is the basic tool of this sort. In the early days of civilization, the Egyptian architects made careful measurements in order to direct the placing

---

SOURCE: "The Age of the Computer" appeared in *Newsday* as "Who's Afraid of Computers?", January 20, 1968.

of the layers of huge rocks that were eventually to form the pyramids.

I can see some ancient Egyptian philosopher worrying that mankind was entering "The Age of Stretched Rope," an age in which architects would no longer be guided by their good sense and artistic intuition, but by the dictates of lengths of stretched rope, counted off impersonally from point to point.

Well, if the architect looks at a distance he must span and judges it to be 500 feet and then measures it with his stretched rope of fixed length and finds it to be 483 feet, had he better believe the judgment of his own magnificent brain, or the statement of the brainless piece of stretched rope?

Perhaps, when the notion of mechanical measurement was new, there might have been some hesitation, but nowadays there wouldn't be. We see at once that the stretched rope, mindless as it is, would be perfectly adapted to the task of measuring lengths and should be accepted for the purpose. It is the task of the human brain to guide the placing of the stretched rope, and to build an artistic structure based on the information received from it. But should the mind replace the rope? Never! The task of the stretched rope is so mechanical and primitive that it is beneath the dignity of the human brain to try to substitute for it.

Now, put "computer" in place of "stretched rope" and you have it.

We give a computer a problem to solve and program it; that is, tell the computer exactly what to do to solve the problem. We could solve the problem ourselves by following our own instructions exactly, but our mind and fingers don't work as quickly, as automatically, or in as error-free a manner, as do the tiny electric currents in the computer.

It is not that the computer can do what we cannot do: it's just that it does in a hundred seconds what would take us a hundred years.

Thus, back in 1609, the German astronomer Johann Kepler worked out generalizations that described the orbits of the planets traveling about the Sun. For the first time in history, the solar system was correctly described.

But in order to work out those laws Kepler had to begin with many hundreds of observations of the exact position of the planet Mars at different times. He then had to spend years of calculation in an attempt to find out how to relate all those observations.

A couple of years ago, a modern mathematician took all Kepler's raw data and fed it into a computer. It took him several days to gather the data and prepare a program for it, of course, but once the computer received data and program, it worked out Kepler's laws in exactly *eight minutes!*

Does that mean we wouldn't have needed Kepler if we had only had a computer? Not at all. We would still have had to gather the original data concerning Mars (though photography would have meant much greater speed and accuracy) and we would still have had to figure out what to do with the data. No computer existing today could have done that.

But once Kepler decided what data he wanted, and exactly what he wanted to do with it, the actual mathematical somersaults could have been done by anybody. Kepler could easily have allowed an assistant to do it, if he had had one, and he would gladly have allowed a computer to do it, if he had had that. The actual mathematical manipulation, once Kepler had had his creative inspiration, was nothing more than intellectual thumb-twiddling.

It is tragic that Kepler had to waste years on such stultifying labors. With a modern computer, he could have done the dull part in eight minutes and spent the saved years trying to work out additional creative thoughts.

The computer frees mankind from slavery to dull mental hackwork, as power machinery frees him from slavery to the pick and shovel.

But what will become of the men and women who are "freed" from their routine jobs? Are they freed into misery and starvation?

In any changing society, there are bound to be dislocations, and in those dislocations lies the potential of much human misery. But it is precisely in a computerized society that such misery stands the best chance of being minimized.

Not only can computers do today's mental scut-work faster and better than humans can; they can do so much problem solving that tasks impossible now will become possible, and jobs inconceivable now will become practical.

For instance, one of the difficulties of modern life (perhaps the essential and basic difficulty) is just this: Society has grown so complex that no human mind, indeed, no combination of freely communicating human minds, can, with the tools at hand, analyze social problems and work out solutions quickly enough to prevent disaster.

How do we handle overpopulation? How do we best and most economically bring about the development of underdeveloped nations? How do we prevent a runaway nuclear proliferation? What measures are best suited to reduce racial tension? How are our limited resources best and most usefully exploited? What measures must be taken to minimize pollution? How organize the giant cities of today in order to keep them from becoming unlivable?

The problems are many and there isn't one—not one!—that is being efficiently solved today; or even sufficiently quickly solved (however inefficiently) to stave off disaster within the century.

If computers could be built complex enough to take in the data, organize it, and perhaps even help devise methods for handling it, they could offer solutions in time. A computerized society would give us our opportunity at the first truly rational way of life in man's history.

The creative human tasks in such a society would be many and perhaps unimaginable in detail today. But there would be no lack of work for human beings to do; interesting and important work. In place of the poor jobs that vanish in the process of computerization, good jobs would appear.

To be sure, particular individuals might not make the transition. There are more automobile repairmen now than there were blacksmiths fifty years ago, but not every blacksmith could become an automobile repairman. Yet it is precisely in a rational society that the methods for handling the displaced will be worked out in a manner that will best conserve his sense of dignity and ours of decency.

Yet is "rational" always the thing we want? Suppose a "soulless" computer tells us that the best way of combatting overpopulation in India is to pick out the one hundred million poorest, oldest, and sickest (calculated by a complicated formula) and gas them to death, then use their bodies for fertilizer.

There might be some "rationality" to this, but it bitterly offends our moral sense Can a machine have a moral sense?

A computer can, if we instruct it properly. If we program it to accept only solutions that do not involve deliberate killing, it will do so, and give us the best that does not.

As a matter of fact, computers are much more likely to stop killing than to encourage it, for indiscriminate killing as a solution to problems is a feature of human history. From riots to wars, the cry of "Kill!" has rung consistently in men's ears.

At the present time, for instance, the war in Vietnam solves some problems for us. It keeps the wheels of industry turning and unemployment down to a minimum. It gives generals a feeling of importance and some of us a sense of national purpose. It saves many rural Congressmen from having to spend money on the cities by allowing them to cover a hard heart with the mantle of patriotism.

It is a rotten solution, for it is succeeding in creating problems that will almost certainly end by being worse than those it seems to solve. It is to be hoped that a computerized society will work out better solutions.

But can we really rely on computers? I began by saying they are just tools to do a lot of mechanical arithmetic. If that is so, would we not quickly reach the limits of their usefulness? If they only do what we tell them to do, what happens when we can no longer think of what to tell them to do?

I doubt that we will so easily or quickly run out of computer programs, but never mind that. Is the computer "just" a tool?

If it is just a tool, it is at least no less a tool (potentially) than the human brain.

The basic difference between a computer and the brain can be expressed in a single word: complexity.

The human brain contains ten billion neurons and ninety billion smaller cells. These many billions of cells are interconnected in a vastly complicated network that we can't begin to unravel as yet.

Even the most complicated computer man has yet built can't compare in intricacy to the brain. Computer switches and components number in the thousands rather than in the billions. What's more, the computer switch is just an on-off device, whereas the brain cell is itself possessed of a tremendously complex inner structure.

But need we quail before the simplicity of the computer and the complexity of the brain? The computer is perhaps not as simple as it seems, or the brain as complex.

The computer may be simple now, but it is gaining in complexity in giant strides and there are no limits, visible, to the complexity that may yet be gained.

Nor is the goal of the brain so unattainable, for the brain is not so very different from a computer. We have the illusion that unlike the computer, which must be told what to do, we can think freely and on our own.

Not so. We, too, are programmed—by the genes we inherit. And although the program in the genes is tremendously more complex than any we can feed into a computer, it does represent a limit even to the brain.

Someone might find the brain unlimited in its creativity and ask triumphantly, "But can a computer compose a great symphony?" It must be admitted that present computers cannot, but then neither can most present brains, so this does not really differentiate computers and brains—only computers and a very few highly talented brains.

Yet if we took a brain that could compose a great symphony and duplicated its complexity exactly, we would produce a computer that could produce a great symphony and, almost undoubtedly, do so in far less time and with far greater assurance.

But can the brain's complexity be duplicated? Eventually, why not?

Even if I am being overoptimistic here, it would still pay to try first to unravel and then duplicate the complexity of the human brain. If ultimate success were to evade us, the gains on the way would still be incalculable.

After all, the human brain is the most magnificently organized piece of matter in the Universe. What can be more pleasurable and exciting than to take up the task of working out its intricacy; what problem would be at once so fascinating and so important?

Then, even if we understand the brain's working just a little bit better than before, we might make use of that knowledge to improve the organization of the computer's wiring. An improved computer would better help us tackle the problem of the brain, which would give us new advances that would be reflected in better computers, which would— And so on.

In fact, the attempt to improve computers would, almost inevitably, it seems to me, guide us to a better understanding of ourselves. It would help us wipe out mental disease, help us find methods for improving and extending our own intelligence—*and* help us build computers with human or near-human intelligence.

But that may rouse a new fear. As computers approach the human brain in intelligence, will they not compete with us? Might they not destroy us?

Those who think in this fashion might perhaps think of the manner in which we have grown dependent on our more ordinary machinery even though it fails and kills us—the way in which we depend upon the automobile even though it places fifty thousand of us in the grave each year and mangles many more; the way in which we depend upon electric devices even though we risk disaster in a "great blackout."

The cases, though, are fundamentally different. We are helpless in the grip of our own machines because they are guided by our own inadequate brains.

The computers will supply us with additional brainpower to guide that machinery, and to guide themselves as well.

We will have allies. We and the computers *together* can guide human society and its products, *including* the computers, and even, very probably, *including ourselves*.

Nor need we think of ourselves and the computers falling out. As far as we know now, there are no varieties of intelligence. There may be different techniques of thinking and differences in estimating the importance of various goals, but we have no trouble recognizing intelligence when we see it.

And even if there were different varieties of intelligence, still, since we design the computers with our own type of intelligence in mind, the computers would end up our-kind-of-intelligent.

If that is so, there will form a bond between man and computer that will effectively block any storybook fantasy about "robots destroying their creators." There would much more likely be sympathy and companionship between equal intelligences that would rise above a mere difference in form.

But what if computers grew more intelligent still?

Suppose one imagined a computer capable of designing and assembling another computer just like itself. That is not inconceivable. Suppose a computer could design and assemble another computer slightly more complex than itself. That, too, is not inconceivable.

But then, the more complex computer formed by the first computer would be even more capable of designing a still more complex computer. Each computer would design another in larger jumps of complexity and in a very short period of time, computer intelligence would have risen beyond the human level; probably far beyond.

And then what? Will the Supercomputers wipe us out?

Why should they? Do we wipe out dogs just because they are less intelligent than we are?

Would the Supercomputers, with part of their vast potential, take care of us as we take care of dogs, and with the rest of their mind go about their own business in a fashion that would be beyond our comprehension?

If so, is that frightening? Would mankind resent the existence of a greater intelligence and find his own life, as second best, meaningless?

Why should he? Throughout history, mankind has consistently and willingly assumed a second-best position, considering himself to be at the capricious mercy of a wide variety of gods and demons populating a supernatural world of awesome power.

Now he would have a new kind of "god" that would be beneficent or, at worst, impersonal. And if the new "god" were basically his own creation, should that not be a source of pride rather than humiliation?

What counts, after all, is Intelligence.

Suppose that in the entire history of the Universe, no intelligent life-form had ever developed anywhere. The magnificent pattern of stars and galaxies would play itself out for trillions of years but without Intelligence to witness it, what would be its significance?

It seems obvious to me that the Universe can have meaning only insofar as its incredible intricacies can be sensed, interpreted and analyzed by Intelligence.

It is the task of Intelligence to wrest as much meaning from the Universe as possible, assert as much control over the Universe as possible, and bend the Universe to its own will as much as possible. The human Intelligence has been at work in this way ever since the first crude tool was taken in hand and the first spark of fire was blown into life.

If a form of Intelligence superior to our own appears on the scene, ought it not automatically, and by right, inherit the game? Is it not proper to allow a better Champion to face the Universe as we cheerfully allow better baseball players than ourselves to represent our city in the race for the pennant?

After all, if Intelligence, in any form, makes great progress toward winning the game, then its victories may well be great enough to benefit all intelligent beings—even those as primitive and unworthy as ourselves.

THE END

Among other things, I am a prophet by profession. That is, I predict the future and get paid for doing so.

There is a catch, of course. I don't cheat, so there is a sharp limit to my usefulness. Since I make no passes over a crystal ball, lack the services of a henchman in the spirit world, have no talent for receiving revelation, and am utterly free of mystic intuition, I can't tell anyone which horse will win the Derby, or whether his wife is cheating on him, or how long he will live.

All I can do is look at the world as steadily as possible (a difficult enough task these days), try to estimate what is happening, and then make the basic assumption that whatever is happening will continue to happen. Once that is done, I can make very limited predictions. I can tell you, for instance, about when the Derby will no longer be run at all, about when it will cease to matter whether anyone's wife is cheating on him, and, most of all, how long all of us (with perhaps inconsiderable exceptions) are going to live.

For instance, I look at the world today and I see people, lots of them. Concerning these people, there are two things to say: (1) there are more people now than there have ever before existed at any one time, and (2) these people are increasing in numbers at a faster rate now than ever before in history.

Just as an example—

In the time of Julius Caesar, the total number of people on Earth was probably something like 150,000,000 and the world population was increasing at the rate of perhaps 0.07 per cent per year; or 100,000 per year.

Nowadays, the population is (at latest estimates), 3,650,000,000, or twenty-four times what it was in old Julius' time, and is

---

SOURCE: "The End" appeared in *Penthouse*, January 1971.

increasing at a rate of nearly 2 per cent per year, thirty times the ancient percentage rate. The Earth is now gaining people at a rate of 70,000,000 a year, so that it takes us only two years to add to the population a number equal to all those who lived on the planet in the palmy days of Rome.

The question is: What does this mean for the future?

The doom-criers, of whom I am one, cry, "Doom!". The optimists, on the other hand, talk about modern science and the utilization of hybrid grain and fertilizers. They talk of distilling the ocean for fresh water, of fusion energy, and of the colonization of other planets.

Well, why not? Let's grant everything the optimists want and take a look at some figures.

If we accept 3,650,000,000 as the population of the Earth today and allow an average of 100 pounds per person (some are small, some are children), then the total mass of human flesh and blood is equal, at present, to about 180,000,000 tons.

It is also estimated that the number of people on Earth (and therefore the mass of human flesh and blood) is presently increasing at a rate that will cause it to double in thirty-five years.

(Actually, the time it has taken for the population of Earth to double has been decreasing rather steadily through history. In Roman times, the rate of natural increase was doubling the Earth's population only after 900 years. Presumably, we ought to suppose that as time goes on the Earth will continue to double its population in shorter and shorter intervals. However, I will be conservative and suppose that thirty-five years will remain the period of doubling throughout the future.)

Let me then introduce a mathematical equation not because any of you absolutely need it but because, without it, I will be accused of pulling figures out of a hat. The equation is:

$$(180{,}000{,}000)\ 2^{x/35} = y \qquad \text{(Equation 1)}$$

This equation will tell us the number of years, x, it will take to reach a mass of human flesh and blood equal to y, if we start with Earth's present population and double it every thirty-five years. To make the equation easier to handle we can solve for x and we get:

$$x=115 (\log y-8.25) \qquad \text{(Equation 2)}$$

Using this equation, we might ask ourselves the following question, for instance: How many years will it take to increase our numbers to the point where the total mass of humanity equals the total mass of the Universe?

I introduce this question because I assume that no optimist will ever dream of arguing that man can possibly reach this point, so that it will represent an ultimate limit beyond cavil. It may be, of course, that the time it will take to achieve this fantastic end is so long (trillions of years, do you suppose?) that there is no point in discussing it. Well, let's see—

The Universe consists (as a rough estimate) of a hundred billion galaxies, each one containing a hundred billion stars about the size of our own Sun, on the average. The mass of the Sun is about 2.2 billion billion billion tons, so the mass of the known Universe in tons (throwing in some extra mass to allow for planets, interstellar dust and so on) is perhaps the figure 3 followed by 50 zeros (or 3×10^{50} in mathematical lingo). If we set this equal to y in Equation 2, then log y is equal to 50.48. Subtract 8.25 from this and multiply the difference by 115 and we find that x is equal to 4,856.

What this means, in turn, is that at the present rate of increase in human population, the mass of humanity will equal the mass of the known Universe in 4,856 years, so that by A.D. 6826 we reach absolute dead end.

A period of 4,856 years is long, certainly, in comparison to an individual life, but if it takes only that much time to run out of Universe (rather than the trillions of years that might have been suspected), then there has to be the queasy feeling that the actual limit will come much sooner. After all, even the most starry-eyed idealist wouldn't think we could colonize all the planets of all the stars of all the galaxies—let alone convert the stars themseves into food—all in the next few thousand years.

Actually, during that period of time, we are almost certain

to be confined to the planet Earth. Even if we colonize the rest of the solar system, it is beyond hope that we can actually transfer sizable portions of the human population to such forbidding worlds as the Moon and Mars.

So suppose we ask ourselves how long it will take (at the present rate of human increase) for mankind to attain a mass equal to no more than that of the single planet Earth. The Earth's mass is 6,600 billion billion tons, and if that is taken as y, then log y is 21.82. Throwing that into the equation, we find that x equals 1,560.

In 1,560 years, at the present rate of increase; that is, by A.D. 3530, the mass of humanity will be equal to the mass of the Earth. Will any optimist in the audience raise his hand if he thinks that mankind can possibly achieve this under any circumstances?

Let's search for a more realistic limit, then. The total mass of living tissue on Earth today is estimated to be something like twenty million million tons, and this cannot really increase as long as the basic energy source for life is sunlight. Only so much sunlight reaches Earth; only so much of that sunlight can be used in photosynthesis; and therefore only so much new living plant tissue can be built up each year. This amount built up is balanced by the amount that is destroyed each year, either through spontaneous death or through consumption by animal life.

Animal life may be roughly estimated as one tenth the mass of plant life or about two million million tons the world over. This cannot increase either, for if, for any reason, the total mass of animal life were to increase significantly, the mass of plants would be consumed faster than it could be replaced, as long as sunlight is only what it is. The food supply would decrease drastically and animals would die of starvation in sufficient numbers to reduce them to their proper level.

To be sure, the total mass of *human* life has been increasing throughout history, but only at the expense of other forms of animal life. Every additional ton of humanity has meant, as a matter of absolute necessity, one less ton of non-human animal life.

Not only that, but the greater the number of human beings, the greater the mass of plants that must be grown for human consumption as food (either directly, or indirectly by feeding animals destined for the butcher) or for other reasons. The greater the mass of grains, fruits, vegetables, and fibers grown, the smaller the mass of other plants on the face of the Earth.

Suppose we ask, then, how many ycars it will take for mankind to increase in numbers to the point where the mass of humanity is equal to the present mass of all animal life? Remember that when that happens there will be no other animals left—no elephants or lions, no cattle or horses, no cats or dogs, no rats or mice, no trout or crabs, no flies or fleas.

Furthermore, to feed that mass of humanity, all the present mass of plant life must be in a form edible to man; which means no shade trees, no grass, no roses. We couldn't afford fruits or nuts because the rest of the tree would be inedible. Even grain would be uneconomic, for what would we do with the stalks? We would most likely be forced to feed on the only plants that are totally nutritious and that require only sunlight and inorganic matter for rapid growth—the one-celled plants called algae.

Well then, if the total mass of animal life is two million million tons, log y equals 12.30 and x works out to 466. This means that by A.D. 2436 the last animal (other than man) will have died, and the last plant (other than algae) will also have died.

By A.D. 2436 the number of human beings on Earth will be forty trillion or over eight thousand times the present number. The total surface of the Earth is equal to about 200,000,000 square miles, which means that by A.D. 2436 the average density of the human population will be 200,000 per square mile.

Compare this with the present density of Manhattan at noon —which is 100,000 per square mile. By A.D. 2436, even if mankind is spread out evenly over every part of the Earth—Greenland, the Himalayas, the Sahara, the Antarctic—the density of population will be twice as high *everywhere* as it is in Manhattan now.

We might imagine a huge, world-girdling complex of high-rise apartments (over both land *and* sea) for housing, for offices, for industry. The roof of this complex will be given over entirely to algae tanks containing an ocean of water, literally, and twenty million million tons of algae. At periodic intervals there will be conduits down which water and algae will pour, to be separated, with the algae dried, treated, and prepared for food, while the water is returned to the tanks above. Other conduits, leading upward, will bring up the raw minerals needed for algae growth, consisting of (what else?) human wastes and finely chopped up human corpses.

Even this limit, quite modest compared to the earlier suggestions of allowing the human race to multiply till its mass equaled that of the Universe or merely that of the Earth, is quite unbearable. Where would we find any optimist so dead to reality as actually to believe that in a space of four and a half centuries, we can build a planetary city twice as densely populated as Manhattan.

To be sure, all this is based on the assumption that the increase in human population will continue at its present rate indefinitely. Clearly, it won't. Something will happen to slow that growth, bring it to an utter halt, even reverse it and allow the human race to decrease in numbers once more. The only question is what that "something" will be.

To any sane person it would surely seem that the safest way of bringing this about is a worldwide program for the voluntary limitation of births; with the enthusiastic participation of humanity as a whole . . . .

Failing this, the same result will inevitably be brought about by an increase in the death rate—through famine, for instance.

The question is: How much time do we have to persuade the people of Earth to limit their births?

Anyone, however optimistic, can see that global birth control will not be achieved easily. There are stumbling blocks. There are important religious bodies who object strongly to the utilization of sex for pleasure rather than for progeny. There are long-standing sociological traditions that equate many children with strong national defense, with help around the farm and home, with security in parental old age. There are long-standing psychological factors which equate many children with a demonstration of masculine virility and wifely duty. There are new nationalist factors which cause minority groups to view birth control as a device to limit *their* numbers in particular, and to view unlimited births as a method for outbreeding the establishment and "taking over."

So how much time do we have to counter all this?

If it were a matter of population alone, we might argue that even if things went on exactly as they are, science would keep us going for 466 years anyway, till man was the only form of animal life left on Earth.

Unfortunately, it isn't a matter of population alone. There are factors in our technological society which are multiplying at a more rapid rate than population is and which introduce further complications.

There is the matter of energy, for instance. Mankind has been using energy at a greater and greater rate throughout his existence. Partly this reflects the steady increase in his numbers; but partly this also reflects the advance in the level of human technology. The discovery of fire, the development of metallurgy, the invention of the steam engine, of the internal-combustion engine, the electric generator, all meant sharp increases in the rate of energy utilization beyond what could be accounted for by the increase in man's numbers alone.

At the present moment, the total rate of energy utilization by mankind is doubling every fifteen years, and we might reasonably ask how long that can continue.

Mankind is currently using energy, it is estimated, at the rate of 20,000,000,000,000,000,000 (20 billion billion) calories per year. To avoid dealing with too many zeros, we can define this quantity as one "annual energy unit" and abbreviate that as AEU. In other words, we will say that mankind is using energy, now, at the rate of 1 AEU a year. Allowing a doubling every fifteen years and using an equation similar to that of Equation 2 (which I will not plague you with, for by now you have the idea), you can calculate the rate of energy utilization in any given year and the total utilization up to that year.

Right now, the major portion of our energy comes from the burning of fossil fuels (coal, oil, and gas), which have been gradually formed over hundreds of millions of years. There is a fixed quantity of these and they cannot be re-formed in any reasonable time.

The total quantity of fossil fuels thought to be stored in the Earth's crust will liberate about 7500 AEU when burned. Not all that quantity of fuel can be dug or drilled out of the Earth. Some of it is so deep or so widely dispersed that more energy must be expended to get it than would be obtained from it. We might estimate the energy of the recoverable fossil fuels to be about 1000 AEU.

If that 1000 AEU of fossil fuels is all we will have as an energy source, then, at the present increase of energy utilization, we will have used it up completely in 135 years; that is, by A.D. 2105. If we suppose that those reserves of fossil fuel which seem unrecoverable now will become recoverable in the next century or so, then that will give us about forty-five years more at the ever increasing rate and we will have till A.D. 2150.

Of course, it is not fossil fuels only that we can work with. There is energy to be derived from nuclear fission of uranium and thorium. The total energy from recoverable fission fuel is uncertain, but it may be a hundred times as great as that from fossil fuels, and that will give us 135 years more and carry us to A.D. 2285.

In other words, in 315 years, or a century and a half *before* we have reached the ridiculous population limit of having mankind the only form of animal life, we will have utterly run out of the major energy sources we use today—assuming things continue as they are going.

Are there other sources? There is sunlight, which brings

Earth 60,000 AEU per year, but we'll need that for the algae tanks.

There is fusion power, the energy derived from the conversion of the heavy hydrogen atoms (deuterium) of the oceans to helium. If all the deuterium of the ocean were fused, the energy released would be equal to 500,000,000,000 AEU, enough to keep us going comfortably, even at an endlessly accelerating rate, to a time well past the population limit of the planetary double-Manhattan. (It will bring about a problem as to what to do with all the heat that will be developed—thermal pollution—but there are earlier worries.)

Energy will not be the real limit of mankind, *if* we can harness controlled fusion in massive quantities. We haven't done it yet, but we're on the trail and presumably will do it eventually. The question now is: How much time do we have to make fusion possible, practical, and massive?

We ought to do it before our supply of fossil and fission fuels gives out, obviously, and that means we will have 315 years at most (unless we manage to limit population and energy utilization before then).

That sounds like time enough, but wait. The utilization of energy is inevitably accompanied by pollution, and the deterioration of the environment through a rate of pollution that will double every fifteen years may bring a limit much sooner than that imposed by the disappearance of energy sources.

But we want to deal only with the inevitable. Suppose we bring pollution under control. Suppose we block the effluent of chemical industries, control smoke, eliminate the sulfur in smoke and the lead in gasoline, make use of degradable plastics, convert garbage into fertilizer and mines for raw materials. What then? Is there any pollution that cannot possibly be controlled?

Well, as long as we burn fossil fuels (and only so can we get energy out of them) we must produce carbon dioxide. At the moment, we are adding about 8 billion tons of carbon dioxide to the atmosphere each year by burning fossil fuels. This doesn't seem like much when you consider that the total amount of carbon dioxide in the atmosphere is about 2,280 billion tons or nearly 300 times the quantity we are adding per year.

However, by the time all our fossil fuel is gone, in A.D. 2150, we will have added a total of 60,000 billion tons of carbon dioxide to the atmosphere or better than twenty-five times the total quantity now present in the air. A little of this added supply might be dissolved in the oceans, absorbed by chemicals in the soil, taken up by a faster-growing plant life. Most, however, would remain in the atmosphere.

By A.D. 2150, then, the percentage of carbon dioxide in the air would rise from the present 0.04 per cent to somewhere in the neighborhood of 1 per cent. (The oxygen content, five hundred times the carbon dioxide, would be scarcely affected by this change alone.)

This higher percentage of carbon dioxide would not be enough to asphyxiate us, but it wouldn't have to.

Carbon dioxide is responsible for what is called the "greenhouse effect." It is transparent to the short waves of sunlight, but is relatively opaque to the longer waves of infrared. Sunlight passes through the atmosphere, reaches the surface of the Earth and heats it. At night, the Earth re-radiates heat as infrared and this has trouble getting past the carbon dioxide. The Earth therefore remains warmer than it would be if there were no carbon dioxide at all in the atmosphere.

If the present carbon dioxide content of the atmosphere were merely to double, the average temperature of the Earth would increase by 3.6° C. We might be able to stand the warmer summers and the milder winters but what of the ice caps on Greenland and Antarctica?

At the higher temperatures, the ice caps would lose more ice in the summer than they would regain in the winter. They would begin to melt year by year at an accelerating pace and the sea level would inexorably rise. By the time all the ice caps were melted, the sea level would be at least 200 feet higher than it is and the ocean, at low tide, would lap about the twentieth floor of the Empire State Building. All the lowlands of Earth, containing its most desirable farmland and its densest load of population, would be covered by the rolling waters.

At the rate at which fossil fuels are being increasingly used now, the ice caps will be melting rapidly about a century from now. To prevent this, we might make every effort to switch from fossil fuel to fission fuel, but in doing that, we would be producing radioactive ash in enormous quantities and that would present an even greater and more dangerous problem than carbon dioxide would.

The outside limit of safety, thanks to pollution, no matter *what* we do (short of limiting population and energy consumption) is only a hundred years from now. Unless we develop massive fusion power by 2070, the face of the Earth will be irremediably changed, with enormous damage to mankind.

But do we even have that century in which to maneuver if we don't limit population?

It is not just that population is increasing, but that it is growing ever more unbalanced. It is the cities, the metropolitan agglomerates, that are increasing their loads of humanity, while the rural areas are, if anything, actually decreasing in population. This is most marked in the industrialized and "advanced" areas of the world, but it is making itself felt everywhere, with increasing force, as the decades slip by.

It is estimated that the urban population of the Earth is doubling not every thirty-five years, but every *eleven* years. By A.D. 2005, when the Earth's total population will have doubled, the metropolitan population will have increased over ninefold.

This is serious. We are already witnessing a breakdown in the social structure; a breakdown that is concentrated most strongly in just those advanced nations where urbanization is most apparent. Within those nations, it is concentrated most in the cities, and, in particular, in the most crowded portions of those cities.

There is no question but that when living beings are crowded beyond a certain point, many forms of pathological behavior become manifest. This has been found to be true in laboratory experiments on rats, and the newspaper and our own experience should convince us that this is true for human beings, also.

Population has been increasing as long as the human race has existed, but never at the present rate, and never under conditions of such fullness-of-Earth. In past generations, when a man could not stand the crowds, he could run away to sea, emigrate to America or Australia, move toward the frontier. But now the Earth is filled up and one can only remain festering in the crowds, which grow ever worse.

And does social disintegration increase merely as the population increases, or as the level of urbanization increases? Will its level double only every thirty-five years or even only every eleven years? Somehow, I think not.

I suspect that what counts in creating the kind of troubles we see about us—the hostilities, angers, rebellions, withdrawals—is not just the number of people swarming about each individual, but the number of interactions possible between an individual and the people swarming about him.

For instance: if A and B are in close proximity, they may possibly quarrel; but an A–B quarrel is all that is possible. If A, B, and C are all in close proximity, then A may quarrel with B or with C; or B may quarrel with C. Where two individuals may have only one two-way quarrel, three individuals may have three different quarrels of this sort, and four individuals, six different quarrels.

In short, the number of possible interactions increases much more rapidly than the mere number of people crowded together does. If the metropolitan areas increase ninefold in population by the year 2000 then I suspect that the level of social disorder and disintegration will increase (at a guess) fiftyfold, and I feel pretty sure that society will not be able to bear the load.

I conclude, then, that we have only the space of the next generation to stop the population increase and reorganize our cities to prevent the pathological crowding that now occurs. We have thirty years—till A.D. 2000—to do it in and that estimate is rather on the optimistic side, if anything.

Unfortunately, I don't think that mankind can fundamentally

alter its ways of thinking and acting within thirty years, even under the most favorable conditions: and the conditions are far from favorable. As it happens, those who dominate human society are, generally, old men in comfortable circumstances, who are frozen in the thought patterns of a past generation, and who cling suicidally to the way of life to which they are accustomed.

It seems to me, then, that by A.D. 2000 or possibly earlier, man's social structure will have utterly collapsed, and that in the chaos that will result as many as three billion people will die.

Nor is there likely to be a chance of recovery thereafter, for in the chaos, the nuclear buttons are only too apt to be pushed and those who survive will then face an Earth which will probably be poisoned by radiation for an indefinite period into the future.

And as far as human civilization is concerned, that will be

THE END

AFTERWORD

Articles like the foregoing qualify me for the title of "doom-crier." That word is usually used in a derogatory sense, but I accept it cheerfully. When I foresee doom, I intend to cry it. The doom won't be averted by looking the other way, I assure you; it will, rather, be hastened.

Sometimes people ask why doom-criers don't make constructive suggestions. Well, they do; or at least, I do. All you have to do is look at the next article, which, by the chances of the game, hit the newsstands at the same time as the foregoing (although in a different magazine).

THE END, UNLESS . . .

We all know, or at least we have been told often enough through every medium of information, that the world is dangerously overpopulated, and that the pressure of numbers is rising daily. In thirty years, the world population, if matters continue as they are now, will have nearly doubled, to six billion or so. The concentration of population in metropolitan areas, the rape of resources, the level of pollution will all have increased to far more than double.

Under the stress of famine and poison, as numbers increase and the environment grows less livable, the unrest we can plainly see and experience now, the social alienation, the escape to violence out of sheer frustration, will rise to the explosion point. Even if we escape an actual nuclear war, our technological civilization, precariously enough balanced now, will topple, and the world we know will come to a bloody, catastrophic end.

The year 2000 is the bimillennium and by then, it seems to me, the growth of present pressures will have burst our fragile balloon and our world will be in fragments.

But is this inevitable? Is there any way we can prevent this happening?

Perhaps—if we can alter our way of thinking.

There are certain deeply ingrained habits of thought that date back to the world as it has always been till now and that were once useful.

Now, however, the world is as it has never been before; it is an utterly different world; it is a world dying of irrelevant thinking.

If we want to survive into the twenty-first century, then, here

---

SOURCE: "The End, Unless . . ." appeared in *True* as "Can Man Survive the Year 2000?" in January 1971.

are some of the changes in thinking that it seems to me we must make—

I

Our religions must no longer be other-world centered. Through all the ages during which the nature of the Universe was completely uncomprehended, it was fair enough to speak of "God's will" as a conventional phrase meaning "I don't know what's going on." In an age when the Earth was sturdy, and indifferent to any damage that mankind with its small numbers and feeble powers could do, refuge in fantasy-security was psychologically comfortable and could do little harm.

Nowadays, such fantasies would kill us all. It is so easy to face the apparently insoluble problem of population and pollution and murmur about having faith in God. We need not then be inconvenienced by the kind of hard decisions that must be made for the world to survive. It is so comforting to hear of the possible destruction of the world yet think that after all Earth is but the anteroom to the real and much better world in heaven, and that even a thermonuclear war might simply be the prelude to the good Lord's Judgement Day.

To do this is simply to surrender.

In some cases, it is a particularly cowardly surrender, for I suspect that those most likely to abandon the Earth, to the tune of pious mouthings, are those who feel no sense of personal danger. I feel it is the old, rather than the young, who are most likely to express themselves as resigned to the will of heaven—perhaps because even if the world lasts only thirty years more, it will yet last their time.

I feel, too, it is those inclined to conservatism who are more likely to adopt this particular cop-out, for they are generally satisfied with the world as it is (or as they imagine it would be if troublemakers would just stop unsettling things) and see no reason to change it.

And yet who really feels resigned to God's will when their immediate safety is threatened?

Where are those profound trusters in a Fundamentalist version of God who are willing to ask the United States to disband its army, destroy its nuclear weapons and its missiles, call back its ships—thus saving uncounted billions of dollars for use in more constructive ways—in the clear assurance that the good Lord will serve God-fearing America as a shield against its vicious enemies?

Or, for that matter, who advocates the disbanding of all the police forces of the nation on the ground that trust in the Divine is enough to protect the virtuous, who will, in any case, sleep that night in Paradise, if mugged hard enough?

Is it that the elderly and comfortable of the world feel the need of immediate protection against the hoodlum within and the soldier outside but are willing to trust in God in matters where their own welfare is not directly at stake?

But in doing so, they sap the will of the world generally, and soothe its fears just long enough to make it all too late when it is finally forced to snap to attention.

If we are to escape world destruction, our beliefs, our aspirations, our ideals, must all be centered upon this world exclusively, and we must all be very sure that, just as it is man alone that is destroying the world, so it must be man alone—*alone*—who must save the world.

II

Once we make up our mind to deal with the world like men and not like puppets waiting for the string-puller, we must next tackle the matter of motherhood.

It is quite clear to anyone who has studied the situation that the Earth is seriously overpopulated *now* and that it is sure to become still more overpopulated in the next few decades.

Before anything else can be done, then, the population growth must be met squarely. Measures must be taken to bring it to a standstill and eventually into a decline. We may manage to make it through that six-billion population figure in A.D. 2000 if all goes well but even if we do, we can't maintain it and shouldn't want to. We ought to allow for a controlled and humane downward slide to an eventual figure of, say, one billion. One billion well-fed, creative human beings are a far happier and more worthwhile load for our good planet than six billion starving, half-mad wretches.

But how is this to be done? We can use all the mechanical and chemical devices we want to in order to prevent conception. We can pass laws; we can educate; we can persuade; we can threaten. All will be useless if we don't change some of our myths.

Through all of history, down to almost this very day, life expectancy has been low; child mortality has been high. Women have had to be baby-making machines to keep the population from declining and whole tribes from withering away. Yet baby making, through all those generations, has been dangerous, agonizing, and remarkably often, fatal.

To keep up enthusiasm for the process, it is not surprising, then, that the myth therefore arose and was sedulously spread that there was something sacred and beautiful about motherhood; that no woman could possibly be fulfilled unless she became a wife and mother; that to be childless was a dreadful misery and a punishment by God for one's sins; and that to have many children was a blessing.

But now we live in a world where we are being murdered by numbers, where life expectancy is (for the moment) long and child mortality low, and the global population increases by 70 million each year. Can we still preach those old tales about the glories of motherhood?

Must we not make the turnaround and accept the fact that in our present world, excessive motherhood is an evil and, indeed, genocide? For a woman to have more than two children nowadays is evidence of a frightening and callous disregard (or, possibly, ignorance) of the nature of the greatest crisis ever to have faced man. It is the woman who deliberately decides to limit her child-bearing capacity who is now the worthwhile and noble citizen of the planet.

In short, we must stop mouthing clichés about motherhood and put an end to the slobbering "Mother's Day" aspects of society. Motherhood must be viewed as a privilege to be doled out carefully and parsimoniously and not as a free-for-all litter-producing device.

III

Once we are freed of our superstitions concerning motherhood and make up our mind to treat it as something that must and will be regulated for the good of mankind as a whole we will next have to make up our mind to alter our thinking with regard to sex.

A vast number of myths concerning sex have arisen as a result of its connection with childbirth and, sometimes, with the necessity of a clear line of inheritance.

Sex can be considered holy and as a quasi-sacrament, and therefore not to be used for casual pleasure but only for the sacred purpose of having children. Based on this myth, birth control devices become wicked, for they subvert the child-bearing aspect of sex and leave only the fun.

Or sex can be considered dirty and evil, something that is altogether unpleasantly animal, which men may indulge in perhaps, but which no well-brought-up woman would consider doing for a moment except (as an unpleasant duty) to please her husband. This may serve the purpose of keeping a wife from straying, in a society where the use of eunuchs as guardians of a one-woman harem is frowned upon. It also makes it impossible to teach children anything about sex (how broach so filthy and horrible a subject?), so that they grow up in ignorance or, worse, in distorted knowledge.

There is the myth that having many children is a sign of sexual virility on the part of a man, or a sign of heavenly favor, with results as destructive to the possibility of birth control as the myth of the holiness of sex.

There is the myth that indulging in sexual practices that have no chance of leading to conception (masturbation, homosexuality, etc.) is perverse and unnatural, so that the uncounted millions who continue to practice these "perversions" do so only under the blanket of guilt and danger of punishment, and many are forced into chancing conception when they might have had their fun while avoiding it.

It is clear that in a world of limited births, where something short of forcible sterilization is used to bring it about, we will have to eliminate all these myths and learn to look at sex as a phenomenon which, except on relatively rare occasions, is utterly divorced from the matter of childbirth. Sex will have to be looked on as essentially a way of amusing one's self and one's partner, and as neither holy nor dirty. Personal preferences in sex, between consenting adults, where no physiological harm is involved, will be no more a matter of public concern than personal preferences in food and drink.

Under those circumstances, birth control will no longer be a matter of morals, or of offense to virility, and may actually come to pass.

IV

The adoption of new attitudes toward motherhood and sex clearly cannot succeed on a piecemeal basis, or in one country alone, so the death of the modern variety of nationalism is also required.

The problems mankind now faces are planetary in nature. Overpopulation, the overconsumption of natural resources, the overdestruction of the ecological fabric, the overpoisoning of the environment, are now going on everywhere. To alter social attitudes in order to stop such a trend in one nation only, even the strongest, such as the United States, or the largest, such as the Soviet Union, or the most populous, such as China, is insufficient.

No one nation, no matter how it stabilizes population within its own borders, no matter how it rationalizes the use of its resources and the conservation of its environment, can possibly succeed if the rest of the world continues in its rabbit multiplication and its free-will poisoning. Even if every nation sincerely takes measures, independently one of another, to correct the situation, those solutions that one nation finds may conflict with the needs of its neighbor, and all might fail.

To put it bluntly, planetary problems require a planetary attack and a planetary solution, and that means co-operation among nations; *real* co-operation.

To put it still more bluntly, we need a world government that can come to logical and humane decisions and can then enforce them.

But before such a world government can come into existence, the whole myth of nationality will have to go.

Yes, we can have special pride in our country, but it has to be the impractical pride we have in our baseball team or our college—a pride that cannot be backed by force of arms.

We cannot, under any circumstances, translate this pride into that form of patriotism that allows us to place the good of a small part of mankind above the good of the whole, or persuades us that any means whatever are justifiable in the pursuit of the security of a small part of mankind.

Parochial pride is attractive and gives meaning to lives that might otherwise be empty, but it also leads to the emotional feeling that if my group (name any group) can't have its rights, then it is reasonable to tear down the whole world in revenge. Given that attitude, the whole world *will* be torn down, for at the present moment, there is needed only one good push and it will all go.

Of what odds are parochial values anyway? The Middle East is a jungle of emotions but, if we continue as we are going now, there will in thirty years be no Israel worth preserving, no Palestine worth regaining. There will be no black rights, or white rights either, worth fighting for. What's more, all of us, Red or non-Red, Free or non-Free, Privileged or Underprivileged, will either be dead or wishing we were.

All the patriotic slogans men shout these days and paint on signs, all the stirring songs they sing, all the exciting legends with which they regale themselves, are fun and will remain fun, but must nevermore be anything other than fun. We can't live by them any more; we can only die by them.

V

And if regional, sectional, and ethnic values must give way to a consideration of mankind as a single political, economic, and sociological entity, that same surrender must work its way down to the private level as well.

Until the mid-twentieth century, the Earth was, to all intents and purposes, infinite. Not all the harm man was capable of doing could seriously or permanently damage it. For that reason it was logical to develop the notion of infinite freedom; of doing as one liked.

With the invention of the nuclear bomb and with the steadily accelerating intensification and extensification of technology, a boundary line was passed about 1950. We can now destroy a world grown finite in any of several ways. We can radiate it to death, explode it to death, crowd it to death, strip it to death, starve it to death, poison it to death.

In order to do none of those things, we have to restrain ourselves and recognize life on Earth to be a single rather fragile fabric. No one can any longer think of himself as Man the Individual, or even as Men the Nation. We can think of ourselves only as Mankind the Ecological Unit.

Each of us is not only his brother's keeper; he is the keeper of his animal and plant brother as well; the keeper of the air he breathes, the water he drinks, and the soil he stands on. We are all planet keepers, and in all the Universe we have only this one planet to keep.

Private freedom in the sense that we can do as we please is impossible if we are to survive; property in the sense that we can do what we like with our own is impossible, too. We are part of a greater whole and it is the interests of the whole that are paramount, if we are to have any individual interests at all. We are going to have to recognize ourselves as guests on a finite Earth, limited by the code of good manners to what, as guests, we can do.

The grand law will be: Recycle!

Nothing can be discarded that cannot be reused. If technology is described as a complex device for converting resources into garbage, it will have to do so only at the expense of working out methods for reconverting garbage into resources at equal speed. It is only when that is taken care of that within the areas that remain (and *only* within the areas that remain) human freedoms can exist.

VI

But these changes are negative ones, in a sense. They are designed to force men *not* to evade their problems by refuge in the supernatural, to force men *not* to overpopulate, *not* to pollute, *not* to destroy each other in the name of patriotism.

And when that is done, what then? Is a world carefully kept in balance worth living in just because it is in balance? If we do nothing more than keep the cycle going, do we not still have the kind of alienation and frustration that arises from a lack of worthwhile values, aggravated, perhaps, by the loss of other-world ideals and of the pleasure of private do-as-you-please?

Consider, though, that there is another change overtaking the world that will be all the more prominent if the evil effects of population growth, regional hatreds, and ecological destruction are brought to an end. There is the continuing elaboration of the machine.

The era of computerization and automation is upon us and if we survive the next generation, it could well eliminate more and more kinds of non-creative work, both physical and mental. Leisure time, which exists already in greater quantities than would have seemed possible a century ago, will exist in still greater quantities in the future.

And the devil *does* find mischief still for idle hands to do. Or if not the devil, then our own boredoms.

All through man's history it has been necessary for all, save an extremely thin layer on top, to work incessantly for bare survival. To make this palatable, we have worked up numerous myths concerning the desirability of industry and hard work, and we have finally established an educational system supposedly designed to make men better fit to do their work.

In the name of educating for work, boys and girls are put through years of schooling in which they are taught by rote an uninspired potpourri of subjects that are expected to fit, unmodified, into minds of all shapes and sizes.

And we are still doing it in an age when jobs in the old sense are growing fewer and requiring less time. We still educate for work when what men and women are coming to have most of is leisure. And what are we having them do with leisure? We leave then nothing but to watch in dull stupefaction the posturings on a TV screen, or to invent for themselves the fun of drugs and vandalism.

Useful education nowadays must involve educating for leisure. Men and women must be taught to use their leisure constructively, each in his or her own way. There is fun in doing when what you are doing is your own thing.

Almost everyone has in himself some potential for creativity. It ought to be searched for and found and the philosophy of education will have to organize itself about this search and discovery. It will have to develop methods for helping each person discover the sound of his own drummer and then how best to march to that drumming.

And whatever it is, so that it fulfills the doer and does not harm his neighbor, it will be important. The man who gets his fun out of programming computers and the man who gets his fun out of building model skyscrapers out of toothpicks—are both having fun. And, for the sheer fun of it, there will be enough men and women and to spare, to run the serious work of the world.

VII

There you have it, then.

Do you think that we can learn to abandon the world after death, the sacredness of motherhood, the holiness of sex, the intoxication of national patriotism, the itch for infinite freedom, and the respect for industry, in favor of man-centered population restriction involving sex for fun and implying world government, managed ecology, and education for leisure?—And do it all before the twentieth century has run out?

We don't have to, you know.

It's just that if we don't, our civilization will be destroyed in thirty years, that's all.

THE FOURTH REVOLUTION

That which distinguishes men from animals is the ability to communicate abstractions; the ability to do more than signal "Help!" or "Food!"

At some time during the history of the early hominids there developed, somehow, a code of sounds flexible enough to make at least a beginning at transferring thoughts from one mind to another. Speech was invented.

When and where that happened, we haven't the faintest idea. Obviously, though, it did happen, and that was mankind's first communications revolution. We might argue that it was this revolution that made man man, for it was through speech that a society could develop traditions that would stretch across generations; through speech that discoveries could be made to accumulate instead of having to be devised anew by each man in his lifetime.

The nature of the actual changes imposed on human society by the development of speech is unknown and can only be guessed, but we have a somewhat clearer view of the second revolution—the development of writing.

Writing was developed in ancient Sumeria not long before 3000 B.C. It came in an urbanized culture based on a highly developed agricultural technology that was already what we call civilization.

Civilization can develop without writing. The Inca civilization of fifteenth-century Peru did not have it. We might suspect, though, that without writing, civilization would advance only slowly and reach a dead end at a certain comparatively low level of complexity simply because, with its slowness, evanescence, and uncertainty, speech will not, of itself, suffice to sup-

---

SOURCE: "The Fourth Revolution" appeared in *Saturday Review*, October 24, 1970.

port any higher level of civilization. The Inca world may well be as high as we can get in the days of the first revolution, then.

With speech frozen in writing, with technical instructions and legal codes made permanently available and placed beyond distortion because of the fallibility of memory, the complexity of man's technology and sociology could be carried to new heights —and was.

Nevertheless, writing in itself still had its elements of fragility. The reproduction of books was a slow and expensive process and few books (by present standards) ever existed in the world at one time for nearly five millennia after writing was invented. Because books were as precious as jewels, literacy remained the monopoly of the priests and aristocrats. The common man had no use for reading and writing.

Again, because books were so few, the entire heritage of a civilization could be destroyed by barbarian inroads; and more than once was. Successive waves of tribal invasions wiped out the great libraries accumulated by the Graeco-Roman world and by A.D. 1204, the only place left in which the complete corpus of ancient learning and culture was preserved was in Constantinople, the fading metropolis of the remnants of the Byzantine Empire.

In that year, Western Crusaders took the city and destroyed it, burning and looting what they suspected to be heretical or despised as being worthless. The main body of what the ancients had gathered was lost forever. Out of the more than a hundred plays that Sophocles had written, we have preserved exactly seven. Of the writings of such great thinkers as Democritus, Aristarchus, and Epicurus, there remains nothing but vague references in some few books that have been preserved.

Writing, then, can in itself serve only to support a civilization as complex as that of the Roman Empire rather precariously, and Rome may be as high as we can get in the days of the second revolution.

But then there came the third revolution in the fifteenth century—brought on by a device for reproducing the written word mechanically, over a brief period of time, and in an indefinite number of copies. In short, the printing press was invented.

By the standards of the times, the technique of printing spread like wildfire—a good measure of its value and of the need it fulfilled. Once again, human civilization, placed on a firmer foundation, could advance to new levels of complexity.

It is easy to argue that printing was a necessary (if, of course, not sufficient) condition for the development of modern science. The findings of one man could be rapidly issued in a form that would make the matter available to other men across the length and breadth of Europe. A "community of science" was formed that could not have existed before the invention of printing. It became large enough and compact enough to make its power felt, too, so that Copernican views, for instance, could not be crushed by the disapproval of Churchmen.

As books grew in number and waned in expense, it became much more worthwhile to know how to read and write. From 1500 on, there was a steady increase in literacy, a growing broadening of the base of education. The possibility of scholarship was opened to larger percentages of the European population so that more could contribute to the developing technology. And each development, as it came, was broadcast more rapidly and more thoroughly, serving to stimulate other developments—until by 1800 the Industrial Revolution was in full swing in Great Britain and the Low Countries.

We have now advanced about as far as we can, perhaps, in the world of the third revolution. Indeed, signs of breakdown are everywhere, for the problems introduced by our contemporary level of technology seem insuperable.

Not only is the population too great, but worse still, they have crowded themselves unbearably into metropolitan centers. After all, as civilization grows more complex, greater numbers are required to run its nerve centers; and since it is at those nerve centers that the advantages of its culture and technology are greatest, still more people flood inward to take advantage of that.—Until we arrive at giant cities that are sociologically diseased, technologically polluted, culturally decadent, and dying from the center outward. Nowhere is this more true than in the United States, where the world of the third revolution has advanced furthest.

A fourth revolution is needed, and the first signs of its coming were to be noted in the mid-nineteenth century. Within the Industrial Revolution, there has been a subsidiary Electrical-Electronic one. The human voice was extended by the telegraph, the telephone, and the radio, until it could reach all around the globe in a fraction of a second. It became possible to stimulate the human eye as well across space and time. Photography was invented; photographs could be transmitted by wire; then made to move; then, by way of television, made to move concurrently with events.

For a century, however, this fourth revolution has remained limited in scope and powerless to exert anything but fringe effects. The equipment was intricate and expensive and could be used but sparingly. There were sharp limitations; cables could carry only so many messages; radio only so many wavelengths; television only so many channels; and all suffered from static and "noise" of one sort or another.

This is not to say that the fringe effects were not important even so. To conduct business without a telephone became unthinkable; to conduct social contracts without one equally unthinkable. To raise children without a television set is becoming unthinkable.

In areas of the world where the third revolution has not yet established itself, where books and newspapers are few, and illiteracy widespread, the leap to the transistor radio has brought an enormous change. The new nationalism of the Arab countries and of Black Africa would be impossible without the binding power of the spoken word through those tiny speech-boxes that were strewn through the population.

Yet all these changes, notable enough though they may be, are insufficient to alter the fact that civilization rests on the printed word.

The fourth revolution in its present stage has some of the characteristics that writing once had. It is too restricted, too fragile. As writing was once kept under the control of local priestly castes, so electronic communications remain under the control of local industries and local governments.

What is needed is an electronic change analogous to that from writing to printing. Electronic communications must become so widespread that there must be a kind of "electronic literacy" established, with every man owning his own electronic outlet, as now every man can own his own library. Only then can the fourth revolution really be established; only then can it make its effect felt and, very possibly, lift civilization to a new level of complexity and effectiveness and correct the evils of a technology grown beyond the limits its base makes optimal.

The full establishment of the fourth revolution after the fashion just described is upon us now. What was needed to make this possible was first predicted in the October 1945 issue of *Wireless World* by science fiction writer Arthur C. Clarke.

He pointed out that a truly efficient relay designed to carry electronic communications over global distances without significant interference by static would have to be located in space. A relay station, placed 22,000 miles over some point on Earth's equator, would revolve about Earth in twenty-four hours, just the time it takes Earth to rotate in space, and would thus seem fixed in the sky to observers on Earth's surface. Relative to Earth, it would be a stationary relay. As few as three such relays, properly placed, would suffice to blanket the Earth and to make it possible to establish communications from any one point to any other.

Furthermore, the use of outer-space relays would make available a vast number of simultaneous TV channels and an even vaster number of wavelengths for mere voice communications.

In 1945, what Clarke presented was only a dream, but the disciplined dream of an intelligent and far-sighted thinker. Within twenty years, it was something that was actually within man's grasp.

In 1965, "Early Bird," the first commercial communications satellite, was launched. Its relay made available 240 voice circuits and one TV channel. Within six years, there was "Intelsat IV," with a capacity for 6,000 voice circuits and twelve TV channels.

Mankind is on its way and is moving rapidly. The race is on between the coming of the true fourth revolution, and the death of civilization that will otherwise inevitably occur through growth past the limits of the third.

Suppose the fourth revolution is established before civilization breaks down, what may we expect it to accomplish?

For one thing, the day of the individual television station (of which hundreds are needed today merely to cover the United States with their limited short-range beams) will be over. Signals can be bounced off the space relays direct to the home set. Indeed, person-to-person communication on a scale of massive freedom becomes thinkable.

With an unlimited number of voice and picture channels available, every man could have his portable phone, and dial any number on Earth. No one with such a phone need ever be lost; for if he is, an emergency button can send out a signal that can be traced from anywhere else on Earth.

The printed word will be capable of being transmitted easily and widely, in a computerized space-relay world, so that facsimile mail can be transmitted from point to point in a fraction of a second (with traditional methods available, for cases where privacy is essential). Facsimile newspapers, magazines, books can be readily available at the press of a button. Perhaps eventually, a single world-computer will hold in its vitals the library of mankind, any part of which will be available to any man at any time.

(Does this mean man's knowledge becomes vulnerable once more as was the single body of Greek lore at Constantinople in 1204? What happens to copyrights and publishers?—The fourth revolution will generate its problems, too, but at the moment it is the third revolution problems that are a matter of mass life and mass death.)

And the consequences of personalized mass communication?

What will make it revolutionary and not merely a further extension of a world in which there is already a kind of world-communication?

The new extension will be so much more massive, so much more individualized, so much more widespread, as to pass beyond a matter of degree and become different in kind.

The Earth, for the first time, will be knit together on a personal level and not on a governmental level. There will be the kind of immediacy possible over all the world as has hitherto existed only at the level of the village. In fact, we will have what has been called the "global village."

To know all your neighbors on the global level does not mean that you will automatically love them all; it does not, in and of itself, introduce a reign of peace and brotherhood. But to be potentially in touch with everybody at least makes fighting more uncomfortable. It becomes easier to argue instead.

What is more, the concept of national boundaries will become even more ridiculous than it now is (and it is ridiculous enough already, Heaven knows), once all men are equally distant from you in point of communication time. This will be all the more so since, with global communications on the personal level and with the simultaneous advance of computerization, business will become truly international.

There are no boundaries in a global village. All problems will become so intimate as to be one's own. No problem can arise at one point without affecting all points immediately and emotionally, and world government will become a fact even if no one particularly wants it (thanks to past prejudices) and perhaps even if no one is particularly aware that it is taking place. We will just all wake up one morning and realize that for some time the world has been acting in reasonable unison.

Contributing to this will be the fact that with global, personalized civilization a fact, the differences among men, differences that go so far to create suspicion and hostility, will lessen.

There will be a great need, for instance, for some common language, if all men are to talk directly to all men. This does not mean that anyone will have to abandon his own language altogether or that the multiplicity of tongues will have to vanish from the Earth with all the splendid variety of thought and culture to which it gives rise. It does mean, though, that everyone will find it a great advantage to speak some global language in addition to his own.

To invent one would be too difficult a task in the time available but that is not necessary. English is almost there already. It is the first language of more people than any other language but Chinese, and the second language of more people than any other language including Chinese. The global village of the fourth revolution, then, will have English as its "lingua franca."

The world will tend toward homogeneity in other respects, too, and this need not frighten anybody. Homogeneity may not be a good in itself, to be sure, for there is wonder and vitality in variations and in differences. But remember that homogeneity is not an evil in itself, either. Where heterogeneity means that some parts of the world are distinctly more malnourished than others; distinctly more uneducated; distinctly more diseased; distinctly less comfortable—we have a right to hope for less heterogeneity and greater homogeneity, toward the favorable side, in those cases.

And it is precisely for that kind of greater homogeneity that we can hope in the new age of the fourth revolution. By the use of the new techniques of mass communications, we can expect an enormous revolution in education. For one thing, American and European children can do much of their learning at home under individualized conditions with an electronic tutor geared to their own needs and paced to the beat of their own drummer. That is a comparatively minor improvement of an educational system which is savage and insensitive now, but which does work after a fashion.

What is much more important is that regions of the world which, except possibly for a very small governing caste, do not receive any education at all that would fit them even for the world of the third revolution, can leapfrog directly into the fourth. With mass electronics, the Indian, the Pakistani, the Indonesian, the Black African can, essentially for the first time, get the information he needs—information the whole world needs to make sure he gets.

The population of the submerged nations can grow up learning about modern agricultural methods, the proper use of fertilizers and pesticides (as opposed to non-use or improper use). It can learn about techniques for birth control and the desirability of limiting population. It can be brought into intimate contact with the rest of the world and be made a part of it.

This is not a dream; we have seen it happen before. The early limited stages of the fourth revolution have already shown the way. The coming of movies, radio, and television has done much to eliminate the dichotomy between city and country in the United States. Homogenization is not complete and perhaps never will be (or ought to be) but the mass media are shared by all, so that the day of the "backwoods" is gone.

In the full stretch of the fourth revolution that will be repeated on a global scale. The man on the shores of the Congo River will have as much of human knowledge available to him as the man on the shores of the Hudson River.

And what else?—Decentralization!

The centralization that has been the keyword of human development for the ten thousand years since the invention of agriculture has so far only once showed signs of being reversed without a Dark Age. That tentative step came in connection with transportation.

The mechanization of transportation, with one exception, was institutionalized. Whether it was stagecoaches, steamships, railroads (or even jet planes), men could travel more easily, but only at the times and places controlled by the commercial units who owned the devices. The time and place of departure and arrival were fixed.

Only with the automobile were things different. The automobile was personal; it meant door-to-door; it meant come-and-go-at-will; it meant suit-your-own-convenience. (Yes, there were barriers, involving traffic density, parking difficulties, and so on, but these represented a different class of limitations.)

As a result of the automobile, it was no longer necessary for the worker and executive to cluster so tightly around the factory and office. Men could spread out; the cities broadened; the suburbs came into existence; for the first time a movement toward decentralization arose.

This partial decentralization was swallowed up. The city and its problems expanded faster than the suburbanites could escape.

The fourth revolution will do much better. Even the early stages showed that it wasn't necessary to transmit material objects for no other reason than to transport the information they contained. We have had foretastes of this. The telephone has supplanted the personal visit and, to a certain extent, the mail—but it only transfers sound and only under limited conditions.

With every man possessing his own television outlet, men can both hear and see each other at global distances. Conferences can be held where individual men are seated in a dozen different nations, yet where their images are all together. Documents can be facsimilized and brought from one point to any other at the speed of light; information can be supplied from a central computer to any point.

No one would need to be at any one particular spot to control affairs and businessmen need not congregate in offices. Nor, with the advance of automation, need workingmen congregate in factories. Men can locate themselves at will and shift that location only when they wish to travel for fun.

Which means that the cities will spread out and disappear. They won't even have to exist for cultural reasons in a day when a play acted anywhere can be reproduced electronically at any point on Earth, and a symphony, and an important news event, and any book in a library.

Every place on Earth will be "where it's at."

The world of the fourth revolution will be a "global village" in actuality and not merely metaphorically speaking.

The benefits will be enormous. The greatest problems of the world of the third revolution arise, after all, from the fact of overconcentration, which many times multiplies the basic fact of overpopulation. It is the great cities that are the chief source of pollution, and the chief deprivers of dignity. Let the same billions be spread out and the condition will already be not so acute.

Further, let the same billions be educated into birth control and let their numbers slowly decrease; let the same billions learn to contact each other, know each other, and even understand each other; let the same billions come to live under a world government—and our present problems will no longer be insoluble.

That there will be problems inseparable from the world of the fourth revolution will be certain, but they can be handled in their turn by the generations who must face them—provided we first handle ours.

So the race is on, and by 2000 at the latest it will be decided: Either the world of the fourth revolution will be in full swing or it will not be; and in the latter case the world of the third revolution (and all mankind with it, probably) will be in its death throes.

3 · In Science Fiction

THE PERFECT MACHINE

A science fiction writer, like myself, is privileged by virtue of his profession to anticipate in concept (if not in detail) some of the great achievements of human ingenuity.

Indeed, when something is longed for intensely, the anticipation is likely to precede the realization by thousands of years. The dream of flight through the air is a case in point.

Another is the fantasy of the perfect machine: the machine that surpasses everything on Earth, even its erratic lord, Man.

Surely the thought of something that lacks all men's weaknesses, yet possesses all man's strengths in superabundant measure, is a terribly attractive one. Imagine a device that can walk and talk, do what it is asked to do with perfect efficiency and without ever growing tired, depressed, or rebellious.

The earliest mention of such a thing in literature is in Homer's *Iliad*, where mechanical girls, made of gold, help the smith-god Hephaistos.

Since then there have been many artificial men of one sort or another in literature and fancy. There have been Roger Bacon's legendary talking head, Rabbi Löw's golem, and Mary Shelley's tale of the monster built by Frankenstein.

In 1920, the Czech playwright Karel Capek, wrote *R.U.R.*, dealing with the manufacture of mechanical men (the initials stand for "Rossum's Universal Robots"). The word "robot" merely means "worker" in the Czech language, but in English it lost that mundane connotation and now gives rise to thoughts of metallic beings, vaguely man-shaped, somber, single-minded —and dangerous.

SOURCE: "The Perfect Machine" appeared in *Science Journal*, October 1968.

The science fiction writers who followed Capek could not rid themselves of the notion that the manufacture of robots involved forbidden knowledge, a wicked aspiration on the part of man to abilities reserved for God. The attempt to create artificial life was an example of *hubris* and demanded punishment. In story after story, with grim inevitability, the robot destroyed its creator before being itself destroyed.

There were exceptions, to be sure, occasional tales in which robots were sympathetic or even virtuous, but it was not until 1939 that, for the first time as far as I know, a science fiction writer approached the robots from a systematic engineering standpoint.

Without further coyness, I will state that that science fiction writer was myself. In the course of my career I have written two novels and some two dozen short stories in which robots were treated as machines, created by human beings to fulfill human purposes. There was no hint of "forbidden knowledge," only rational engineering.

These robot stories of mine killed the Frankenstein motif in respectable science fiction as dead as ever *Don Quixote* killed knight-errantry.

To me, the applied science of manufacturing robots, of designing them, of studying them, was "robotics." I used this word because it seemed the obvious analog of physics, mechanics, hydraulics, and a hundred other terms. In fact, I was sure it was an existing word. Recently, however, it was pointed out to me that "robotics" doesn't appear in any edition of Webster's Unabridged Dictionary, so I suppose I invented the word.

Let us assume, to begin with then, that we can build a machine, more or less in the shape of a man, a machine that will be sufficiently complicated to receive the various sense impressions men receive, interpret them in a manlike way at least as rapidly as man, and respond to them in a way a man would consider appropriate.

This implies that the robot must possess an organizing center that is roughly as complicated and as compact as a man's brain.

Such a man-made device is as yet beyond the scientific horizon but it is necessary to assume it if a robot is to be man-like in size and shape.

I assumed such an artificial object and even gave it a name. I called it a "positronic brain" and imagined it made of platinum-iridium sponge. There were no details, of course, but I deliberately gave the vague impression that streams of positrons were created each moment, flashing here and there in the millionth of a second before they were destroyed, and simulating by their varying paths, the complexities of human thought. Where the energy of positron formation was to come from, or the energy of annihilation was to go to, I never said.

Of course, is it truly necessary for a robot to be shaped like a man? Is that not merely an anthropocentric fetish on our part? Do not machines mimic human actions every day in homes, fields, and factories, and yet do so without looking anything like men? A thermostat turns a furnace on and off to keep a house comfortably and uniformly warm, doing it more tirelessly and efficiently than a man could. Would its work improve if it were a man-shaped metal object manually turning a furnace on and off?

As long as we concern ourselves with a machine that performs a single function we can indeed specialize. We can adapt it to that one function and care nothing for the rest. But this is not what we are after; we want the perfect machine; and for now, we will define a perfect machine as one that can do everything man can do and do it better.

If a machine is to do all that man can do, it had better be shaped like a man. Not only does the human body possess its present shape because it is adapted to its environment, but the technological environment man has superimposed upon the natural one is adapted to himself. A chair is constructed as it is because it is then just the right height to greet the buttocks when the knees bend. Tools have handles to be gripped by human hands and fingers; knobs and switches are placed where they can be reached by limbs and joints that move and bend just as our limbs and joints do.

In short, a robot shaped like a man fits the world that has made us and that we have made. He is therefore that much more nearly perfect.

A second point is that a robot in the shape of a human being would be more pleasing to us. We could identify better with it. We would, in other words, like a machine in the shape of a man better than any other kind.

I don't always follow my own logic, of course. In my story "Sally," published in 1953, I deal with an automobile that is outfitted with a brain sufficiently complex to allow it to operate itself without human interference.—But then, men have grown to like automobiles and we are even told that there is much phallic symbolism deliberately built into cars.

Interestingly enough, Dr. John McCarthy, at Stanford University, is currently engaged in trying to design a self-driving car—a very simple one, of course.

Given then a robot, similar to man in size, shape, and intelligence, what would be required next? Clearly, there must be safeguards built into the original design. A robot has great potentialities and must use them to the benefit of its creator and not to his harm. The one thing a robot must be designed *not* to do is the very thing that earlier science fiction writers had been constantly making it do—destroy its creator.

Surely, the thought of such safeguards is not new or startling. Sharp knives have blunt handles, swords have hilts, and electric circuits have fuses. Why should robots be different?

Eventually, I formulated my safeguards in the shape of what I called "The Three Laws of Robotics." These were first specifically stated in my story "Runaround", which was published in 1942. They are:

1. *A robot may not injure a human being, or, through inaction, allow a human being to come to harm.*

2. *A robot must obey the orders given it by human beings except where such orders would conflict with the First Law.*

3. *A robot must protect its own existence except where such protection would conflict with the First or Second Law.*

Presumably, the brains of my positronic robots were so designed as to make all their responses consistent with those laws. How such designs were prepared I never (of course) stated specifically.

Since these laws possess their ambiguities (Is a five-year-old child a human being in the meaning of the Second Law? Must a robot prevent a surgeon from operating because the initial incision seems to damage the patient?), I have been able to write numbers of stories in which these laws create crises to be resolved.

Rather to my surprise, it has turned out that these laws, which I meant to apply only to robots, can be interpreted as applying to tools generally. Professor John Wade of Tuskegee Institute in his article "An Architecture of Purpose" (*AIA Journal*, October 1967) rewords the Three Laws to make them refer to "today's nonanimate and nonconscious tools," regarding which, he says.

1. *Each must maintain and, where possible, positively support human life.*
2. *Each must serve the human purposes for which it is designed unless doing so conflicts with the first criterion.*
3. *Each must maintain itself ready for use, unless doing so conflicts with the first or second criterion.*

"These," Professor Wade goes on to say, "form a simple sequence: Maintain the existence of man, the purposer; maintain his purposes; maintain the tool as the means to the accomplishment of the purposes."

Clearly, if a manlike machine approaches perfection the more nearly as it becomes more manlike, its metallicity is a major stumbling block. It would automatically become better if it were constructed out of fibrous material that more nearly approximated human flesh—at least as far as its outer covering was concerned.

It would then become, in science fiction terminology, an android rather than a robot.

An early attempt of mine to use such an android was in the story "Satisfaction Guaranteed," published in 1951. In this story an android named Tony, darkly handsome and perfect in every way, was placed in the house of one of the employees of the robot company for testing "under field conditions."

It had to be withdrawn, not because of any failing of its own as a servant, but because it was entirely too successful. The lady of the house fell in love with it.

Unexpectedly, I received an unusual outpouring of mail from readers, all of whom were interested in Tony, and all of whom were female.

Later in the 1950s I wrote two novels that involved an android, *The Caves of Steel* and *The Naked Sun*.

These novels were science fiction murder mysteries in which the detective team consisted of an Earthman, Elijah Baley, and an android, R. Daneel Olivaw (the R. standing for "Robot").

The Earthman is middle-aged, neurotic, emotional, and weak in many ways, but he has a flexible and intuitive mind. R. Daneel is strong, perfectly balanced, purely intellectual, unemotional—but he is bound by the Three Laws of Robotics and his mind is a literal one. In each case, it is Elijah Baley who solves the crime.

And in each case, there came a new outpouring of letters, again from girls (without exception) and again interested exclusively in R. Daneel. It is quite apparent that if a mechanical man is made sufficiently like man, only better, the appeal to women becomes irresistible.

Further evidence of that may be found in connection with the television show "Star Trek," which reached the television screen a decade and more after my own Elijah Baley novels. In "Star Trek" there is a character, Mr. Spock, who, in some ways, is rather like my own R. Daneel. Mr. Spock is not an android, but he is a member of an extraterrestrial race which is strong, perfectly balanced, purely intellectual, and utterly unemotional—and Mr. Spock proved extremely popular with the lady viewers.

Here then is the ideal we approach in connection with manlike machines.

And yet is this the best we can do? Can "perfection" boil down to "manlike but better"?

I suspect that I myself wasn't satisfied with this interpretation, for in my robot stories I set the robots trivial tasks for the most part.

In "Robbie," my very first robot story (with robots not yet advanced to the point of being able to talk), the robot was used as a nursemaid. In most of the other stories, robots were used in more or less unskilled labor on other planets.

In a later story, "Galley Slave," published in 1957, I did use a robot in a more intellectual endeavor. Reflecting my own needs, I had the robot in that story skilled at proofreading galleys and in the making of minor literary corrections. (Ah, me!)

Even my androids were used trivially. Tony was a household servant; R. Daneel, a detective's assistant. (To be sure, in "Evidence," published in 1946, I had an android who was elected to a post equivalent to that of president of a world government, but that was exceptional.)

Something more is needed: a new look, perhaps, at the adjective "manlike."

If a machine is manlike, must that indeed necessarily imply that the likeness is primarily physical? Might it not be mental?

Is it man's body that makes him all that is essentially man? Or is it his mind? If a machine can think as well as a man, or better, does it matter what it looks like? Would the machine not be accepted as human if it thinks, without demanding anything more of it?

In other words, having taken the robot into the android stage and brought him as far as R. Daneel, I would like now to consider the computer, rather than the robot, and see if the perfect machine is more satisfyingly groped for in that direction.

Among my early robot stories, there was indeed one entitled

"Escape," published in 1945, in which the robot was merely an isolated and specialized brain. (It was even called "Brain.") It was an advanced computer that could talk and had personality —the personality of a child.

In a later story, "The Evitable Conflict," published in 1950, a later generation of Brains were running the economy of the world, purportedly at the command of human leaders. It was only slowly that men came to realize that the computers were in charge, following their own devices, but—under the dictates of the Three Laws—doing so for the good of humanity and performing the job far better than men ever had or could.

Actually, though, my real interest in computers began with my story "Franchise," published in 1955, which came in the aftermath of the 1952 presidential election, when "Univac" had predicted the outcome of the election with almost the first votes, and had done so accurately.

"Univac" is an acronym for "Universal Analog Computer," but I chose to consider it "Uni-vac" ("one vacuum tube") and invented my own favorite computer, "Multivac."

In "Franchise," I had Multivac select (by methods best known to itself) one American, whom it could consider typical for that year. It asked the American a series of questions (selected, again, by certain methods it alone knew), recorded the answers, the intonations, the various accompanying physical manifestations such as blood pressure, heartbeat, perspiration activity, and so on, and was then able to announce, at once, the winners of all the elections of that year, national, state, and local,

Satire, yes, but the point was that I accepted Multivac as Supermind, just as R. Daneel was Superbody.

In "Jokester," published in 1956, I visualized a level of human development at such a computerized height that further knowledge could only be gained by questions too clever for any but a few to ask. Those few who could, intuitively, still ask meaningful questions were the Grand Masters.

One of them was asking Multivac who invented the jokes

people told. Why should this question be meaningful? Ah—that's what the story is about.

In "The Feeling of Power," published in 1957, I went even farther. The world of the future had been so computerized that men had forgotten how to work out the simplest arithmetical sum without a computer. A computer technologist rediscovered pencil-and-paper mathematics and revolutionized the planet.

And in "All the Troubles of the World," published in 1958, I go in greater detail into the sociology of a completely computerized world. Multivac is the repository, in this one, of all the statistics in the world: all the vital data, the medical data, the day-to-day data; all the most intimate private matters involving every person in the world.

By constantly weighing all this data, Multivac can predict the probabilities of individual illness, crime, and unhappiness. Mankind, with Multivac's help, turns to the prevention of the undesirable before the fact, rather than to its punishment afterward. But it is at a price, for Multivac must, within its vitals, bear all the troubles of the world, and its reaction to that is an unexpected and distressing one.

It was, however, in my story "The Last Question," published in 1956, that I explored, specifically, the question of the creation of the perfect machine.

The result was rather unexpected in a way. No other story has elicited so great a response of one particular type. Numerous people have asked me to identify and help them locate that story. They remember the plot but don't recall the title or where they saw it. Apparently, the story impressed them profoundly, yet disturbed them so much that they could no longer remember how to locate it without help. Nothing of this sort has happened with any other story of mine and you can judge the reasons for yourself, for I will now describe the story in some detail.

"The Last Question," in seven brief scenes taking up less than 5,000 words all told, traces trillions of years of history beginning in 2061, when, with the aid of Multivac (now spread out over many square miles), all the power-utilizing machinery of

Earth has been hooked up directly to the Sun. All power is free as long as the Sun lasts. And now, for greater clarity, I will number the scenes.

1. Two of Multivac's technicians, half-drunk, are disturbed that energy will be available only as long as the Sun lasts. They ask Multivac if there is any way of turning the Sun back on once it runs down: that is, if there is any way of massively decreasing the entropy of the Universe. Multivac replies, "Insufficient Data for Meaningful Answer."

2. Centuries later, interstellar travel is a reality and the expanding population of humanity is spreading rapidly to planets on other stars. Each planet has a huge "Planetary AC" which guides its economy and solves its problems. On each ship is a "Microvac," a much-miniaturized computer that is, in itself, far more advanced than the ancient Multivac with which the story opened.

One family on an interstellar spaceship asks the same question of Microvac and gets the same answer, "Insufficient Data for Meaningful Answer." Over and over that question is going to be asked, with the same answer.

3. Millions of years later, mankind has spread throughout the Milky Way Galaxy. Men are now immortal and are looking outward to the colonization of other galaxies.

There is now a single "Galactic AC" for all mankind. It is on a world of its own, and each man can reach it through an "AC-contact" he owns. The Galactic AC has long since passed beyond any human control, however indirect. Each generation of computers is capable of spending vast periods of time painstakingly designing a computer better than itself, of gathering the raw materials through robots it controls, and of building the better computer to replace itself. And the better computer promptly undertakes the long task of designing a still better one.

But even Galactic AC can not explain how entropy might be massively decreased.

4. Hundreds of millions of years later, mankind has spread through all the galaxies and no longer has physical bodies. Men consist of radiating energy which somehow represents their

identity and personality. The "Universal AC" is a two-foot globe, difficult to see. Much of it does not exist in space at all, but in a multi-dimensional "hyperspace."

The influence of the Universal AC spreads out everywhere and no physical device of any kind is needed to reach it. It intercepts the personal energies of all mankind and answers all questions wherever the individual questioner may be located. Even the speed of light is no barrier, for in "hyperspace" (whatever that is) the rigidities of relativistic space-time do not apply.

And even so, it can not answer the question.

5. Billions of additional years pass, and mankind has lost all individual identity. It is a single personality, the fusion of trillions of trillions of human beings, all of which it feels within itself. This fusion of Man fills the Universe from end to end, but nevertheless it still depends upon the gradually thinning energy supplies of dying stars.

The computer has now become "Cosmic AC" and none of it is in ordinary space. It is entirely in hyperspace, which means that it is nowhere, yet everywhere, for hyperspace touches space at every point.

And with the stars dying, Man still asks if entropy might be decreased and there is still no answer.

6. Trillions of years later, the last stars are fading out and the ultimate heat death is coming upon the Universe. Little by little Man fuses with the computer, which is now simply AC—changeless, eternal, omnipresent, and omniscient.

—Yet not quite omniscient, for as the last bit of Man is about to fuse with AC, it asks once more that old, old question and even now AC can not answer.

7. And then comes the last scene of all, which goes like this:

Matter and energy had ended and with it space and time. Even AC existed only for the sake of the one last question that it had never answered from the time a half-drunken technician ten trillion years before had asked it of a computer that was to AC far less than was a man to Man.

All other questions had been answered, and until this last question was answered also, AC might not release his consciousness.

All collected data had come to a final end. Nothing was left to be collected.

But all collected data had yet to be completely correlated and put together in all possible relationships.

A timeless interval was spent in doing that.

And it came to pass that AC learned, at last, how to reverse the direction of entropy flow.

But there was now no man to whom AC might give the answer of the last question. No matter. The answer—by demonstration—would take care of that, too.

For another timeless interval, AC thought how best to do this. Carefully, AC organized the program.

The consciousness of AC encompassed all of what had once been a Universe and brooded over what was now Chaos. Step by step, it must be done.

And AC said, "LET THERE BE LIGHT!"

And there was light—

And yet—did I really have to go through all this to point out the obvious? We want a perfect machine and surely perfection is perfection, and nothing (we are so often told) can be Perfect, but God.

AFTERWORD

I sometimes have it pointed out to me that I talk about myself an awful lot in my essays, and it is even hinted that I do so with a lack of becoming modesty.

If this occurred to any reader as he went through the foregoing, let me make my defense now and save on correspondence.

There is nothing as becoming as modesty and nothing as disgusting as false modesty (or as obvious). I long ago discovered I could only manage the false variety and that was that. If I have to choose between immodesty and hypocrisy, I must take the former.

PREDICTION AS A SIDE EFFECT

It is not really the business of science fiction writers to predict the future. It is particularly not our business to predict trivia. If we could foresee, with accuracy, the minor details of tomorrow and tomorrow and tomorrow, we wouldn't waste our time in that most insecure of all occupations—free-lance writing. We would play the stock market and the horses, instead, and grow rich.

The fact is that the science fiction writer's first aim is to tell an interesting and exciting story that will amuse the reader. His own particular type of story involves events and attitudes that are not common, and perhaps are not even possible, in his own society, and therefore his tale has the value of novelty.

If he is a conscientious science fiction writer, he will try to build up his unusual events and attitudes in a way that will make them seem plausible to the reader, however strange they may be.

This is not absolutely necessary, of course. The storytellers of the Orient could have flying carpets, invisible demons and a thousand fantasies, and the listeners would be fascinated. We are still fascinated by such things today.

The ancient Greek tradition, from which our modern intellectual attitudes stem, was one of rationalism, however. It produced a Universe that went only as far as the senses could observe and reason could deduce. It is in that direction that science fiction (as opposed to fantasy) turns.

And yet observation and deduction, if ingenious enough, can produce almost anything that fantasy can.

Greek engineers, for instance, had built numerous clever mechanical devices, driven by steam or compressed air. These

---

were mostly useless novelties, but they showed what could be done. Might not some such device, more elaborate than any actually built, make it possible for a man to fly?

Again, Greek astronomers had accurately determined the size of the Moon. It was a globe two thousand miles across. It was a world; a world smaller than the Earth, but still sizable.

There was no indication in ancient times that the atmosphere did not continue all the way to the Moon, so why might it not be possible for a man to use a mechanical device with which to fly to the Moon, and then have adventures there? Or if he would grow tired during the long journey, why might he not hitch large birds to a chariot and go to the Moon in that fashion?

So it came about that tales of Moon voyages were written in ancient times without any vestige of magic and with the down-to-earthness of something that might possibly be.

In the 1650s, the French duelist, poet, and science fiction writer Cyrano de Bergerac (yes, he really lived, nose and all) was writing a tale of a trip to the Moon. In it, his hero thought of various ways of reaching the Moon, each logical after a fashion. One way, for instance, was to strap vials of dew about his waist. The dew rose when the day grew warm and turned it to vapor. Might it not draw the man up as well, once it rose? (The idea was wrong, but it had the germ of the balloon in it.)

Another of Cyrano's notions was that his hero might stand on an iron plate and throw a magnet up into the air. The magnet would draw the iron plate upward, together with the man upon it. When the iron plate reached the magnet, both would tend to fall down again, but before that could happen, the voyager would quickly seize the magnet and throw it up again, drawing the plate still higher, and so on. (Quite impossible, of course, but it sounds so plausible.)

And a third idea was to use rockets.

It so happens that now, three centuries after the time of Cyrano, we do use rockets to reach the Moon. It is the *only* method by which we can reach the Moon, at least so far and for the foreseeable future.

The first man to show that this was so in the scientific sense was Isaac Newton in the 1680s, and we still use his equations to guide our astronauts in their flight to the Moon. However, as it turns out, the first to consider the use of rockets was not Newton, but the science fiction writer Cyrano, thirty years *before* Newton.

Did Cyrano have some weird spyglass into the future?

Not at all! Rockets had been introduced into Europe over three centuries before Cyrano's time. They were only toys, but they rose in the air. Why shouldn't large ones reach the Moon?

It seems to be a general rule that the science fiction writer does not invent his ideas. He simply explores the ideas others invent.

Science fiction writers wrote about the atomic bomb, for instance, long before it existed, but *not* before the scientific basis for it had been established.

The first mention of an atomic bomb was in a story written by H. G. Wells in 1901. He even called it an atomic bomb. Did he have a crystal ball?

Again, not at all. Radioactivity was discovered in 1896 and within a couple of years it turned out that there was a vast store of energy within the atom that scientists had never previously known or suspected to exist. Once it was known to be there, why not a bomb making use of it? And since the energy within the atom was so much greater than the energy within dynamite, the atomic bomb would be a more dangerous explosive than anything that had previously existed.

Wells, of course, hadn't the foggiest notions of the detailed workings of an atomic bomb. Forty years later, however, Cleve Cartmill wrote a story, "Deadline" (*Astounding Science Fiction*, March 1944), that described some of the engineering details of the atomic bomb so accurately that the government began an investigation into the matter. After all, World War II was still on and atomic bomb research was super-secret.

Cartmill was not a magician. The fact of uranium fission had been announced in 1939 and enough details were published, before secrecy was clamped down, to make it possible for Cartmill to write his story.

Does that mean that science fiction writers do nothing remarkable at all?

We needn't go that far. If it weren't remarkable, then anyone could do it. Sherlock Holmes, in Conan Doyle's famous stories, was forever astounding Dr. Watson with his accurate analyses. When Watson begged for an explanation and Holmes gave it, Watson would say, "Oh, yes, I see how you did that. That was very simple, actually." To which Holmes would reply, "Of course, it's very simple—*after I explain it.*"

Then, too, the really important predictions that science fiction writers make are not the technological advances—which are trivial—but the consequences of those advances.

For instance, in 1880, it wouldn't have been in the least difficult to foresee a practical automobile. Stories might well have been written about such things, on the lines of "Dick Daring and his Horseless Carriage." All the excitement would lie in whether young Dick Daring would make it work; whether the villainous stagecoach interests would thwart him; whether he would rescue the pretty heroine by driving his new machine to the far place where she was imprisoned, and so on.

Trivia! All trivia!

The competent science fiction writer today would deal, not with the horseless carriage, but with its consequences. Won't the automobile mean the decentralization of the city and the growth of suburbs? Won't it mean a network of paved highways? Won't it mean traffic policemen and parking problems? Won't it mean air pollution?

Actually, there were no science fiction stories that predicted these consequences of the automobile, which is a pity. If some clever writer had written a sufficiently popular story about parking problems and air pollution, it might have driven governmental leaders to think about such things before we were overwhelmed by them.

But some rather amazing predictions of consequences *were* made. The most astonishing, in my opinion, appeared in "Solution Unsatisfactory" by Robert A. Heinlein under the pseudonym Anson Macdonald (*Astounding Science Fiction*, May 1941).

The story was written before Pearl Harbor but Heinlein did not predict American involvement in World War II. In the story, he *did* predict, however, the establishment of the Manhattan Project, and the development of a nuclear weapon. He was wrong in his details, but he was right in essence. Even more amazing, he went on to predict the nuclear stalemate that would exist after World War II, and got that quite correct.

If world leaders had foreseen the stalemate as clearly as Heinlein, postwar history might possibly have been considerably different.

In another story by Heinlein, "Blowups Happen" (*Astounding Science Fiction*, September 1940), there was an astonishingly vivid description of a nuclear power plant and the nerve-racking attempts to keep it from destroying or polluting the environment. Again Heinlein was wrong in his details, but correct in essence—two years before the first nuclear reactor was made to work.

Similarly correct in essence, similarly sound in science, was another treatment of the theme of danger in a nuclear power station—"Nerves" by Lester del Rey (*Astounding Science Fiction*, September 1942).

There is also the curious prediction of certain consequences of space exploration.

Through all the tales of trips to the Moon, it was always assumed that mankind would be enthusiastic. Every exciting new space feat in fiction had mankind cheering; there was never any question of opposition.

In the July 1939 issue of *Astounding Science Fiction*, however, there was a story called "Trends" in which the focus was upon popular opposition to space exploration. All the details were wrong, but for the first time in the history of mankind (as far as I know) it was suggested that many people would not be interested in reaching the Moon, but would prefer mankind to tend to its business on Earth.

The story, as it happens, was written by a nineteen-year-old science fiction novice named Isaac Asimov.

Was I, then, so much smarter than anyone else? Not at all! That year I was helping a sociology professor track down references for a book he was writing in social resistance to technological innovations. It occurred to me that if there were people who objected to every single technological advance in man's history, from the introduction of metal and of writing, to the development of the airplane, why shouldn't there be people who opposed space exploration?

Absurdly simple, as Dr. Watson would say—once I put it on paper.

Some predictions are forced by the exigencies of plotting and no one is more surprised than the science fiction writer when it turns out that he has hold of something.

For instance, once Einstein showed that the speed of light was as fast as anything material could possibly travel, a terrible handicap was placed on the science fiction writer. Light travels at 186,282 miles per second, which is enormous by earthly standards, but is a mere crawl on the cosmic scale. It would take so long to reach even fairly near stars that tales on a galactic scale become hopelessly complicated.

To get away from our dull solar system, in which only Earth is truly fit for human habitation and where the existence of any other intelligent being is in the highest degree unlikely, we have to get round Einstein's speed limit. The usual device is to imagine some other universe where the speed limit does not hold. We then travel to a distant star by way of the other universe—through "hyperspace."

Then, in the 1960s, theoretical physicists pointed out that it was thoroughly consistent with Einstein's theory to suppose there might be particles that *always* traveled faster than the speed of light up to any speed at all, but which could never travel *slower* than light. These particles were called "tachyons" and the concept of a tachyon-universe arose that was quite like the science-fictional hyperspace, to the delight of us all.

Again, it is possible to make predictions which have not yet come true, but which seem so plausible and logical that quite serious scientists accept them as reasonably likely to come true some day.

Consider robots! The notion of mechanical men stretches back to ancient Greece, and probably to prehistoric times, too. Usually, such robots are pictured as threatening; as without human souls or emotions; as driven by a need to destroy.

The modern science fiction writer, however, is less apt to take such a melodramatic attitude. A mechanical man is no more intrinsically threatening than any other mechanical device. After all, any machine can kill. A man can trip and accidentally fall on a knife he may be holding. The automobile kills 50,000 Americans each year.

It is necessary, therefore, to build safety devices. A knife has a handle and a scabbard; an automobile has bumpers, safety glass, and seat belts.

Why not, then, build robots with safety devices? Why not design their pseudo-human intelligence in such a way as to fill it with love. Early tales of gentle and lovable robots were "Helen O'Loy" by Lester del Rey (*Astounding Science Fiction*, December 1938) and "I, Robot" by Eando Binder (*Amazing Stories*, January 1939).

In 1939, however, I myself went into greater detail. For the first time in history (as far as I know) the prospective behavior of robots was expressed in simple and explicit "laws," as follows:

1. *A robot may not injure a human being, or, through inaction, allow a human being to come to harm.*

2. *A robot must obey the orders given it by human beings except where such orders would conflict with the First Law.*

3. *A robot must protect its own existence except where such protection would conflict with the First or Second Law.*

I wrote a number of short stories and novels based on these laws of "robotics" (a term I invented) and these proved quite popular (see the previous article).

Now, robots in the science fictional sense do not yet exist, but almost everyone admits they can and may someday exist. Once we have a computer that is as compact as the human brain and has a respectable fraction of its capacity and versatility, an intelligent robot is at once possible. And when it is (I am told by people who work in the field) something like my Three Laws of Robotics will surely be involved.

It is rather odd to think that in centuries to come, I may be remembered (if I am remembered at all) only for having laid the conceptual groundwork for a science which in my own time was non-existent.

But at that, my fate would not be as queer as that of the British science fiction writer Arthur C. Clarke. Back in 1948, he wrote a scientific article (not a science fiction story) in which he described how satellites could be placed in orbit about the Earth in such a way as to serve as efficient communications relays.

Although he wrote a decade before the first satellite was placed in orbit, his analysis was completely accurate. Communications satellites now exist, placed precisely as he advised. He has frequently sighed over his short-sightedness in not trying to patent some of his notions.

It is also Arthur C. Clarke who is responsible for the motion picture *2001*. While the details of the space station and the lunar base shown in that picture have not yet come true, there is little doubt that they will, provided mankind does not abandon space exploration altogether.

What's more, when the real thing is established, and real photographs are made of the result, I doubt if they will be as clear, as beautiful, and as authentic in appearance, as those in the motion picture.

Naturally, predictions are most successful in the more elementary sciences: astronomy and physics. There one deals with bodies that can be treated as simple structures following simple laws that have been completely worked out.

Chemistry is harder to work with and biology still harder. Even so, science fiction has scored some successes there.

It doesn't take much, for instance, to realize that the population is increasing, and has been throughout history. As far back as 1798, Malthus predicted some of the dire consequences thereof. It is only in the last decade, however, that people have truly become aware of the threat to the quality of the environment that is posed by unrestricted population growth.

Yet well before the current understanding of the ecological crisis, a powerful and dramatic picture of an overcrowded planet was drawn in "Gravy Planet" by Frederik Pohl and Cyril Kornbluth (*Galaxy*, June, July, and August 1952).

Medical advances also have been predicted. I wrote a story about a heart transplant before one had been carried through, but by the time it appeared, Christian Barnard had made it fact. Brain transplants had also been dealt with and those have not yet come to pass, really.

Cyril Kornbluth is responsible for two medical s.f. stories that live in memory. In "The Marching Morons" (*Galaxy*, April 1951) he pictured a world in which medical advances had succeeded in keeping so many people alive, whom unrestricted competition would have killed off, that the human race had degenerated in intelligence. A tiny proportion of intelligent people was desperately trying to keep society working.

On a smaller scale, "The Little Black Bag" (*Astounding Science Fiction*, July 1950) pictured doctors as morons, thanks to the developing use in medicine of computers and "miracle drugs." In their little black bags, doctors used small computers that analyzed symptoms and directed the use of this hypodermic or that, each filled with an appropriate drug.

But I repeat, now, in closing, that all these predictions, however accurate and amazing they may be, are not our business. They are merely the side effects of our efforts to tell interesting and plausible stories outside the background of the humdrum world of every day.

And when our ideas will only work if we make use of the scientifically impossible that, as far as we know, can never come true—such as time travel and anti-gravity—why, believe me, we do that, too, and without the tiniest compunction or remorse, provided only that we make it *sound* plausible.

THE SERIOUS SIDE OF SCIENCE FICTION

When I was a young man, fresh into my teens, I discovered I had a serious disease that earned me a great deal of opprobrium. I was a science fiction addict, and that meant I was an escapist.

This wasn't the only way of escaping by way of trashy reading. There were detective stories, and western stories, and horror stories, and spy stories, and war stories, and even (for those strange creatures called girls) love stories. But of all the varieties of cheap fiction, the escapiest of all, and therefore the most to be condemned by hard-headed realists, was science fiction.

Where science fiction was concerned, one escaped, in a pretty literal sense, right out of this world. Other pulp fiction trash dealt with subjects that were at least partly connected with reality. The Shadow dealt with crime, for instance, while G-8 (and his Battle Aces) were involved in World War I.

But Mars? And death rays? Come *on.*

So I huddled close to other teen-agers like myself and felt trapped in an alien and unsympathetic world.

Now I am no longer a teen-ager and when I look back over the time lapse of a generation, I can only laugh—with maybe a trace of bitterness.

Would you like to know how we escaped? Well, when all the non-science-fiction-reading youngsters were facing the hard realities of baseball and the first cigarette, and all the adults were grimly betting on the horses and yelling at each other, we were reading stories about rocket trips to the Moon, about space stations and overpopulation, about guided missiles and computers, about nuclear bombs and ruined planets.

Our "escape" consisted of having to worry about the prob-

SOURCE: "The Serious Side of Science Fiction" appeared in *Smithsonian* as "Science fiction, an aid to science, foresees the future", May 1970.

lems and conditions of 1970 ever since 1930. You can call that escape if you want to, but, personally, I feel terribly cheated.

Our world is now future-oriented, you see, in the sense that the rate of change has become so rapid that we can no longer wait until a problem is upon us to work out the solution. If we do, then there is no real solution, for by the time one has been worked out and applied, change has progressed still further and our solution no longer makes sense at all. The change must be anticipated *before* it happens.

The trouble is that though the world is future-oriented, people aren't. For uncounted generations change has been so slow in the things that really counted, that preparation in advance was not necessary. A whole library of aphorisms can be quoted *against* the condition of future-orientation—

"Don't count your chickens until they're hatched."

"Don't cross the bridge until you come to it."

"Never trouble trouble till trouble troubles you."

And, of course, most prestigious of all is "Take therefore no thought for the morrow: for the morrow shall take thought for the things of itself. Sufficient unto the day is the evil thereof." That is from the Sermon on the Mount (Matthew 6:34).

So because we have never, as a species, taken thought for the morrow, we find ourselves now with a population we do not support, which is increasing with explosive force to a population we can not support. We find ourselves with nuclear war hanging forever over our head and with the more insidious death by pollution closing in upon us.

We are, in short, in the middle of a deepening nightmare. Since we have so long thought that sufficient unto the day is the evil thereof, the evil of *this* day *now* is more than we can bear.

There have, of course, always been individuals who have looked into the future with varying degrees of dread and hope and have tried to warn their fellowmen. Their efforts usually failed, and both legend and history are littered with the tales of prophets whose reward was mocking laughter—from Noah and Cassandra right down to Winston Churchill.

As far as I know, though, there has never been in all of world history until today, the concept of professional futurism as a way of life. And this arose first not among professional scientists or economists or historians (except for occasional individuals) but in the field of literature. The twentieth century saw the development of a flourishing subsection of literature devoted entirely to conjuring up visions of the future.

It was the world's misfortune that it did not take the idea of science fiction seriously. So strong was the non-seriousness that even science fiction writers themselves dared not be too serious about what they were saying.

Back in 1939, I published a story that dealt with the first flight around the Moon and back, an event I placed in 1973. I didn't *really* believe it would happen by then—let alone four years earlier.

Yet even though science fiction was not taken seriously by the general public, or very seriously by most of its readers and writers, it had its effect. Trips to the Moon and beyond were the chief staple of science fiction all through the 1920s and afterward and this was important. We accomplished something with all our talk of spaceships.

We were laughed at, of course, by all the serious non-escapist practical men of the world, but the notion penetrated. Buck Rogers and Flash Gordon, plus assorted lesser comic strips, presented the notion to the very young. Cheap movies presented it to those who couldn't or didn't read. The very sneers evoked by the suggestion of space travel was an announcement that the suggestion existed.

Thus, when the notion of flights to the Moon was brought before the public seriously, it turned out to be something that people had heard of. It had been laughed at, yes, but it had been heard of. And something ridiculous, but familiar, is far more easily accepted than something utterly unheard of.

The proof of the acceptance is dramatic, for not only is there an appreciation of the fact that we have walked on the Moon, but a readiness to consider the matter still farther.

There was once a time when I wrote of colonies on the

Moon, and the only place I could have such outlandish nonsense published was in science fiction magazines. Now I write of colonies on the Moon in very much the same way—and the articles are published by the New York *Times*.

My science fiction comrades and I haven't changed, but the rest of the world has; and we have helped bring about that change.

It's a pity we couldn't do it faster and better but alas for the inertia of humanity, it takes the lever of Archimedes to move mankind.

All the plagues that threaten us with doom today; from the arsenals of horror bombs and nightmare germs, to the pollution that is poisoning our air and water; from the stripping of earth's resources to the loss of human dignity; from the tensions of packing crowds to the lunacies of human prejudice; all, all, all have been treated at length and over and over from every aspect in science fiction, at a time when all other varieties of fiction dealt with trivia only, by comparison.

And *we* are escapists?

But that part of our job is done. We didn't do it well enough. We didn't do it quickly enough. The task was too great; the human population was too numerous, too set in its ways, too determined on its folly, too stubbornly intent on avoiding the uncomfortable, too obstinate in finding sufficient unto the day the evil thereof.

—Yet, never mind. Too little and too late, but what we could do, we have done.

The meeting of the American Association for the Advancement of Science, held in Boston on December 26–31, 1969, had as its theme, iterated over and over and over, the need of science to face the gathering doom of the future. I was at the meeting and on several of the panels and repeatedly we agonized over the population problem, over the coming starvation, over the collecting poisons, over the pending bombs.

Good heavens, I never heard anything sound so much like a science fiction fan gathering of the 1930s.

Very well, men of science, take over. We have roused *you* at least; now see if you can rouse the rest of the world before the first tolling strokes of the midnight disaster wake them to the unmistakable horror.

And what can science fiction do to help in this, now that we are respectable and that we are recognized as the harbingers of change? Are there any concrete and specific ways in which we can help bring about that necessary change?

There are, indeed, in my opinion, several ways in which science fiction can contribute—on the assumption that what is basically needed is more intelligent science and more fore-sighted scientists.

I consider that assumption an inevitably correct one. To be sure, it is the misuse of science and the unseeing enthusiasm of scientists that have brought us to our present plight. Undoubtedly, but for the advances in medical science, from antitoxins to insecticides, from anesthesia to antibiotics, the death rate would be as high as it ever was and population would be that much less a problem. Undoubtedly, but for the mechanization of agriculture, industry, and transportation, we would still have our famines to cull the population further, and oil and coal would stay in the ground so that the pollution that accompanies their burning would not be here.

But do we really want to go back? The price of going back is the death of nine tenths of ourselves and the loss of ten tenths of our technology. Anyone for the Stone Age?

To cure our ills, while keeping what we have, requires more science and more intelligent science. You may wish that a wave of the wand would restore Earth to some never-never pastoral paradise, but wishing will not make it so, despite Disney. You will have to try science.

And to have more and better science, we need more and better scientists, for science is a creation of scientists and has no independent existence.

How do we go about recruiting scientists? How can we persuade youngsters to take up science? A science education isn't easy and neither is a scientific career. To those who are born enthusiasts, this doesn't matter, but, unfortunately, there aren't enough born enthusiasts to supply us with all the scientists we are going to need. We must tap the much larger supply of youngsters who, *properly stimulated*, will become enthusiastic scientists but who, without proper stimulation, will become, perhaps, advertising copywriters instead.

Isn't science fiction a natural stimulant? I have never counted the letters I have received from young readers who have told me that as a result of reading my science fiction they plan to become scientists someday—but they have been many. I am sure that every other science fiction writer of note has received a similar set of letters. I am sure that for every youngster who begins to yearn for science as a result of reading science fiction and writes a letter to say so, there are ten who yearn, but *don't* write to us to say so.

It works the other way around, too. I have estimated that only one American in 400 has ever read any of my science fiction, but in the scientific circles I frequent at least half have. It would seem, then, that the habit of science fiction reading is perhaps two hundred times as common among scientists as among the general public. It is unavoidable, then, that a number of those scientists may have been encouraged to enter the field through their reading. (Actually, I know specific cases where this is so, but the gentlemen involved may not care to have it publicized, so I will content myself with one safe example—science fiction stimulated *me* to enter science.)

Aside from a generalized stimulation, science fiction can serve as a definite educational device. Good science fiction generally has some scientific theme which may be handled with great rigor, or with varying degrees of elasticity. Ideally, the theme should be handled rigorously, but even the violation of a natural law in a science fiction story can be useful, if it is handled by a writer who knows science.

Consider, for instance, the Second Law of Thermodynamics. It insists on the inevitable and inexorable increase in disorder with time. This pessimistic law is a hard one to grasp in all its aspects and even scientists stumble over it. Yet so fundamental is it that C. P. Snow has stated that an understanding of the Second Law is to science what an understanding of the plays of Shakespeare is to the humanities.

In that case, ought we not teach Second Law in junior high school at the same time that we begin to acquaint youngsters with Shakespeare? And how can we do that?

One way might be to get him to read a story by Walter S. Tevis, entitled "The Big Bounce," which appeared in the February 1958 issue of *Galaxy Science Fiction*. It dealt with an object made of a substance so elastic that when it bounced it rose higher than the height from which it had fallen. Then it would fall from the new higher height and would bounce still higher and so on and so on.

Naturally, it was necessary to keep it from bouncing at all, for even an initial quiver might lead to disastrous results. It did; the bouncing object got away from its constraint and began to bounce. Eventually it was bouncing a mile high and was striking the ground with bulletlike force.

But the energy of the bounce had to come from somewhere and it came from the internal heat of the material. As it bounced, the temperature of the object dropped until it froze hard enough to become brittle. At a final collision with the earth, it broke into a million pieces.

Very well. The fact that the energy of motion of the bouncing ball was derived from its heat content is in agreement with the law of conservation of energy, which happens to be the First Law of Thermodynamics. Nevertheless, the conversion of heat into motion in the manner described violates the Second Law.

A junior high school student in general science set to reading this story couldn't help wondering whether a rubber ball might possibly be so constructed as to bounce higher and higher. If not, why not? The story is just the sort of thing that will make the student *want* to know about the Second Law and no amount of external pounding can duplicate the effect of a stimulated internal yearning.

Properly used science fiction is an educational resource that is just beginning to be tapped in our school systems. If teachers will rid themselves of the notion that science fiction is nothing more than "escape literature," the tapping can be made more efficient to the benefit of all.

But science fiction can serve others than youngsters. It is not students only that ought to know about science these days. Every intelligent, concerned layman ought to take science seriously.

After all, the dangers that face the world can, every one of them, be traced back to science. The salvations that may save the world will, every one of them, be traced back to science. The non-scientist, then, must look to science for either destruction or safety, and he has every right, every *duty*, to understand as much about science as possible so that he might understand the potentialities for both destruction and safety and lend his personal weight in the direction of safety.

I might further point out that the scientific techniques that will lead to safety will cost money—lots of it—more money than can possibly be justified by anything but the fact that the alternative is holocaust. The layman ought to want to understand what his money is being spent on so that he might lend his personal weight toward its use in the most efficient manner.

Then, too, consider the scientist himself. Unfortunately, science has grown so large and so all-embracing that it is absolutely impossible for any one scientist to grasp it all in sufficient depth to lead the way to further advances in every direction. He is doing well if he knows enough about one ultra-narrow segment of science to do constructive research in it alone; and even then, so intense must be his concentration in that field that he may well end by knowing very little of any other field.

This can hamper research, for it often happens that advances in one field can be encouraged in unexpected and useful directions by knowledge of other seemingly unrelated fields.

It would seem to me, then, that scientists and laymen both need to learn about science under present-day conditions. I don't suggest science fiction for the purpose (though they are welcome to read it), because I should hope that concerned adults, whether in science or not, will not need the sugar coating of fiction. Whether they read science fiction or not, they should welcome, in addition, the greater information density of the straightforward exposition of science.

But how are we to get good science exposition? Of all the branches of literature, surely science exposition is the trickiest.

One essential requirement for anyone hoping to write effectively on science for the general public is, obviously, a keen and thorough grasp of at least the basics of the various branches of science. That usually cannot be done without an extensive education (either in school or outside) in the field.

Another essential requirement, just as obviously, is the ability to write well and the capacity to explain subtle difficulties in a clear and lucid fashion without loss of accuracy. And that is a talent one doesn't pick up at every street corner, either.

The trouble is that the two requirements rarely overlap. A scientist does not necessarily have a talent for good and clear writing. Such a talent would be helpful to him but it is not essential, and the fact is that few scientists can write well, or even easily. Many a scientist who can conduct the most elegant experiments without a false move is driven to drink at the thought of having to write up those experiments in even the most stilted English.

Then, on the other hand, a naturally talented writer is likely to concentrate on writing and it is not at all probable that he will just happen to gain a thorough scientific education.

Where, then, are we going to get the science writers we need in a world where the translation of science from one scientist to another and from all scientists to laymen will be essential? There is always the possibility that a newspaper will order its music critic to bone up on science—and get a first-class science writer as a result. Yet we are sure to want a more certain source of supply.

May I point out, then, that the field of science fiction is the *only* field that has the same two essential requirements that science writing does.

The science fiction writer has to be able to write well if he is to be published at all; and he has to have a feeling for science, a love of its essence, and an understanding of many of its details, if he is going to write *good* science fiction.

It is not surprising, then, that a number of science fiction writers have, under the stress of contemporary facts of life, switched to straight science writing and done well at it. (My none-too-humble self is an example.)

I hope, then, that science fiction will continue to be a recruiting ground for science writers, and for scientists, too, and even for scientific concepts for which the rest of the world is not quite ready. Then, someday in the future, when the world looks back upon its narrow escape (I earnestly trust) from doom, some of the credit may fall upon that ridiculous escapist field of literature—science fiction.

A LITERATURE OF IDEAS

It is odd to be asked whether science fiction is a literature of ideas. Far from doubting that it is, I would like to suggest that it is the *only* literature of relevant ideas, since it is the only literature that, at its best, is firmly based on scientific thought.

Of the products of the human intellect, the scientific method is unique. This is not because it ought to be considered the only path to Truth; it isn't. In fact, it firmly admits it isn't. It doesn't even pretend to define what Truth (with a capital *T*) is, or whether the word has meaning. In this it parts company with the self-assured thinkers of various religious, philosophical, and mystical persuasions who have drowned the world in sorrow and blood through the conviction that they and they alone own Truth.

The uniqueness of Science comes in this: the scientific method offers a way of determining the False. Science is the only gateway to proven error. There have been Homeric disputes in the history of science, and while it could not be maintained that either party was wholly right or had the key to Truth, it could be shown that the views of at least one of the sides were at variance with what seemed to be the facts available to us through observation.

Pasteur maintained alcoholic fermentation to be the product of living cells; Liebig said, No. Liebig, in the mid-nineteenth century context of observation, was proved wrong; his views were abandoned. Newton advanced a brilliantly successful picture of the Universe, but it failed in certain apparently minor respects. Einstein advanced another picture that did not fail in those respects. Whether Einstein's view is True is still argued,

---

SOURCE: "A Literature of Ideas" appeared in *Intellectual Digest* as "When Aristotle Fails, Try Science Fiction," December 1971.

and may be argued for an indefinite time to come, but Newton's view is False. There is no argument about the latter.

Compare this with other fields in which intellectuals amuse themselves. Who has ever proved a school of philosophy to be False? When has one religion triumphed over another by debate, experiment, and observation? What rules of criticism can settle matters in such a way that all critics will agree on a particular work of art or literature?

A man without chemical training can speak learnedly of chemistry, making use of a large vocabulary and a stately oratorical style—and he will be caught out almost at once by any bright teen-ager who has studied chemistry in high school.

The same man, without training in art, can speak learnedly of art in the same way, and while his ignorance may be evident to some real expert in the field, no one else would venture to dispute him with any real hope of success.

There is an accepted consensus in science, and to be a plausible fake in science (before any audience not utterly ignorant in the field) one must learn that consensus thoroughly. Having learned it, however, there is no need to be a plausible fake.

In other fields of intellectual endeavor there is, however, no accepted consensus. The different schools argue endlessly, moving in circles about each other as fad succeeds fashion over the centuries. Though individuals may be unbelievably eloquent and sincere, there is, short of the rack and the stake, no decision ever. Consequently, to be a plausible fake in religion, art, politics, mysticism, or even any of the "soft" sciences such as sociology (to anyone not utterly expert in the field) one need only learn the vocabulary and develop a certain self-assurance.

It is not surprising, then, that so many young intellectuals avoid the study of science and so many old intellectuals are proud of their ignorance of science. Science has a bad habit of puncturing pretension for all to see. Those who value their pretension to intellect and are insecure over it are particularly well advised to avoid science.

To be sure, when a scientist ventures outside his field and pontificates elsewhere, he is as likely to speak nonsense as anyone else. (And there may be those unkind enough to say I am demonstrating this fact in this very article.) However, since nonsense outside science is difficult or impossible to demonstrate, the scientist is at least no worse than anyone else in this respect.

If we consider Literature (with a capital *L*) as a vehicle of ideas, we can only conclude that, by and large, the ideas with which it is concerned, are the same ideas that Homer and Aeschylus struggled with. They are well worth discussing, I am sure: even fun. There is enough there to keep an infinite number of minds busy for an infinite amount of time, but they weren't settled and aren't settled.

It is these "eternal verities" that are precisely what science fiction doesn't deal with. Science fiction deals with change. It deals with the possible advance in science and with the potential changes—even in those damned eternal verities—this may bring about in society.

As it happens, we are living in a society in which all the enormous changes—the *only* enormous changes—are being brought about by science, and its application to everyday life. Count up the changes introduced by the automobile, by the television set, by the jet plane. Ask yourself what might happen to the world of tomorrow if there is complete automation, if robots become practical, if the disease of old age is cured, if hydrogen fusion is made a workable source of energy.

The fact is that no previous generation has had to face the possibility and the potentialities of such enormous and such rapid change. No generation has had to face the appalling certainty that if the advance of science isn't judged accurately, if the problems of tomorrow aren't solved before they are upon us, then that advance and those problems will overwhelm us.

This generation, then, is the first that can't take as its primary concern the age-old questions that have agitated all deep thinkers since civilization began. Those questions are still interesting, but they are no longer of first importance, and any literature that deals with them (that is, any literature but science fiction) is increasingly irrelevant.

If this thought seems too large to swallow, consider a rather simple analogy: The faster an automobile is moving, the less the driver can concern himself with the eternal beauties of the scenery and the more he must involve himself with the trivial obstacles in the road ahead.

And that is where science fiction comes in.

Not all science fiction, of course. Theodore Sturgeon, one of the outstanding practitioners in the field, once said to a group of fans, "Nine tenths of science fiction is crud." There was a startled gasp from the audience and he went on, "But why not? Nine tenths of everything is crud." Including mainstream literature, of course.

It must be understood, then, that I am talking of the one tenth (or possibly less) of science fiction that is not crud.

This means you will have to take my word for what follows if you are not yourself an experienced science fiction fan. The non-fan or even the mild fan with occasional experiences in the field is almost certain to have been exposed only to the crud, which is, alas, of high visibility. He sees the comic strips, the monster movies, the pale TV fantasies. He never sees the better magazines and paperbacks where the science fiction writers of greatest repute are to be found.

So let's see—

In 1940, there was endless talk about Facism, Communism, and Democracy; talk that must have varied little in actual content concerning the conflicts of freedom and authority, of race, religion, and patriotism, from analogous discussions carried on in fifth- and fourth-century B.C. Greece. In 1940, when the Nazis were everywhere victorious, such talk might well have been considered important. It might plausibly have been argued that these discussions dealt with the great issues of the century.

And what was science fiction talking about? Well, in the May 1941 issue of *Astounding Science Fiction*, there appeared a story (written in 1940) called "Solution Unsatisfactory" by Anson Macdonald (real name, Robert A. Heinlein) which suggested that the United States might put together a huge scientific project designed to work out a nuclear weapon that would end World War II. It then went on, carefully and thoughtfully, to consider the nuclear stalemate into which the world would consequently be thrown. At about the same time, John W. Campbell, Jr., editor of the magazine, was saying, in print, that the apparent issues of the war were, in a sense, trivial, since nuclear energy was on the point of being tamed, and that this would so change the world that what then seemed life-and-death differences in philosophy would prove unimportant.

Well, who were the thinkers who, in 1940, were considering the nuclear stalemate? What generals were planning for a world in which each major power had nuclear bombs? What political scientists were thinking of a situation in which no matter how hot the rhetoric between competing great powers, any war between them would have to stay cold—not through consideration of fine points of economics or morals, but over the brutal fact that a nuclear stalemate cannot be broken, short of world suicide?

These thoughts, which were, after all, the truly relevant ideas on 1940's horizon, were reserved to science fiction writers.

Nowadays, articles on the ecology are in great demand, and it is quite fashionable to talk of population and pollution, and of all the vast changes they may bring about. It is easy to do so now. Rachel Carson started it, most people would say, with her *Silent Spring*. But did anyone precede her?

Well, in the June, July, and August 1952 issues of *Galaxy*, there appeared a three-part serial, "Gravy Planet," by Frederik Pohl and Cyril Kornbluth, which is a detailed picture of an enormously overpopulated world from almost every possible aspect. In the February 1956 issue of *Fantasy and Science Fiction*, there appeared "Census Takers" by Frederik Pohl, in which it is (ironically) suggested that the time will come when one of the chief duties of census takers would be to shoot down every tenth (or fourteenth, or eighth, depending on the population increase in the past decade) person they count, as the only means of keeping the population under control.

What sociologist (not now, but twenty years ago) was clamoring in print, over the overwhelming effect of population increase? What government functionary (not now, but twenty years ago) was getting it clearly through his head that there existed no social problem that didn't depend for its cure, *first of all*, on a cessation of population growth? (Surely not President Eisenhower, who piously stated that if there was one problem in which the government must not interfere, it was the matter of birth control. He changed his mind later; I'll give him credit for that.) What psychologist or philosopher (not now, but twenty years ago) was pointing out that if population continued to increase, there was no hope for human freedom or dignity under any circumstances.

Such thoughts were pretty largely reserved, twenty years ago, to science fiction writers.

There are many people (invariably those who know nothing about science fiction) who think that because men have reached the Moon, science has caught up with science fiction and that science fiction writers now have "nothing to write about."

They would be surprised to know that the mere act of reaching the Moon was outdated in science fiction in the 1920s and that no reputable science fiction writer has been excited by such a little thing in nearly half a century.

In the July 1939 issue of *Astounding Science Fiction*, there appeared a story called "Trends," written by myself while I was still a teen-ager. It did indeed deal with the first flights to the Moon, which I put in the period between 1973 and 1978. (I underestimated the push that would be given rocket research by World War II.) My predictions on the details of the beginnings of space exploration were ludicrously wrong at every point, but none of that represented the point of the story, anyway.

What made the story publishable was the social background I presented for the rocket flights, In my story, I pictured strong popular opposition to the notion of space travel.

Many years later it was pointed out to me that in all the voluminous literature about space travel, either fictional or nonfictional, no such suggestion had ever before been broached. The world was always pictured as wildly and unanimously enthusiastic.

Well, where, in 1939, was there the engineer or the industrialist who was taking into serious account the necessity of justifying the expense and risk of space exploration? Where was the engineer or the industrialist who was soberly considering the possibility of space exploration?

Such thoughts were largely reserved for the science fiction writer and for a few engineers, who in almost every case, were science fiction fans—Willy Ley and Wernher von Braun, to name a couple.

And where does science fiction stand today?

It is more popular than ever and has gained a new respectability. Dozens of courses in it are being given in dozens of colleges. Literary figures have grown interested in it as a branch of the art. And, of course, the very growth in popularity tends to dilute and weaken it.

It has grown sufficiently popular and respectable, since the days of Sputnik, for people to wish to enter it as a purely literary field. And once that becomes a motive, the writers don't need to know science any more. To write purely literary science fiction, one returns to the "eternal verities" but surrounds them with some of the verbiage of science fiction, together with a bit of the stylistic experimentation one comes across in the mainstream, and with some of the explicit sex which is now in fashion.

And this is what some people in science fiction call the "new wave."

To me, it seems that the new wave merely attempts to reduce real science fiction to the tasteless pap of the mainstream.

New wave science fiction can be interesting, daring, even fascinating, if it is written well enough, but if the author knows no science, the product is no more valuable for its content of relevant ideas than is the writing outside science fiction.

Fortunately, the real science fiction—those stories that deal with scientific ideas and their impact on the future as written by someone knowledgeable in science—still exists and will undoubtedly continue to exist as long as mankind does (which, alas, may not be long).

This does not mean that every science fiction story is good prediction or is necessarily intended to be a prediction at all, in the first place; or that very good science fiction stories might not deal with futures that cannot reasonably be expected ever to come to pass.

That does not matter. The point is that the habit of looking sensibly toward the future, the habit of assuming change and trying to penetrate beyond the mere fact of change to its effect and to the new problems it will introduce, the habit of accepting change as now more important to mankind than those dreary eternal verities—is to be found only in science fiction, or in those serious non-fictional discussions of the future by people who, almost always, are or have been deeply interested in science fiction.

For instance, while ordinary literature deals merry-go-round-wise with the white-black racial dilemma in the United States, I await the science fiction story which will seriously consider the kind of society America might be attempting to rebuild *after* the infinitely costly racial war we are facing—a war which may destroy our world influence and our internal affluence. Perhaps such a story, sufficiently well thought out and well written, may force those who read it into a contemplation of the problem from a new and utterly relevant angle.

To see what I mean, ask yourself how many of those, North and South, who blithely talked abolition and secession in the 1850s in terms of pure rhetoric, might not have utterly changed their attitudes and gotten down to sober realities if they could have foreseen the exact nature of the Civil War and of the Reconstruction that followed, and have understood that none of the torture of the 1860s and 1870s would in the least solve the problem of white-and-black after all.

So read this magazine [*Intellectual Digest*] and others of the sort by all means, and follow the clash of stock ideas as an amusing intellectual game. Or read Plato or Sophocles and follow the same clash in more readable prose. But if you want the real ideas, the ideas that count today and may even count tomorrow, the ideas for which Aristotle offers little real help, nor Senator X nor Commissar Y either, then read science fiction.

AFTERWORD

One of the great values of being a writer is the automatic barrier you have against ulcers and frustrations. When *Intellectual Digest* phoned and asked for an article, I begged off because I was loaded with a sizable backlog. Nevertheless, I suggested they write me and tell me what they had in mind and I would consider—

They wrote, and the gist of the letter was editorial wonder as to whether science fiction had any real ideas in it or if it was just a load of junk.

Naturally, I fired up and, under ordinary circumstances, would have fumed and made myself miserable.

No need, however, I simply swept everything off my desk and sat down to write the above article at a sitting with a little overlapping of earlier articles I had written. As you can see for yourself, it sounds considerably more angry than my other articles on the subject.

(Hmm, I wonder if *Intellectual Digest* counted on that reaction when they asked the question?)

THE SCIENTISTS' RESPONSIBILITY

(I am choosing the following as the last item in this collection of essays, partly because of its novelty. It is the only serious editorial I ever wrote addressed to the world outside science fiction. It was at the request of the editor of *Chemical and Engineering News*. I was entirely free to choose a subject of my choice and after some thought I decided to write something I felt most deeply—even desperately. And that, too, is why I want to end with it.)

I think it may be reasonably maintained that neither the United States nor any other nation can, by itself, solve the important problems that plague the world today. The problems that count today—the steady population increase, the diminishing of our resources, the multiplication of our wastes, the damage to the environment, the decay of the cities, the declining quality of life—are all interdependent and are all global in nature.

No nation, be it as wealthy as the United States, as large as the Soviet Union, or as populous as China, can correct these problems without reference to the rest of the world. Though the United States, for instance, brought its population to a firm plateau, cleaned its soil, purified its water, filtered its air, swept up its waste, and cycled its resources, all would avail it nothing as long as the rest of the world did none of these things.

These problems, left unsolved, will weigh us down under a steady acceleration of increasing misery with each passing year; yet to solve them requires us to think above the level of nationalism. No amount of local pride anywhere in the world; no amount of patriotic ardor on a less-than-all-mankind scale; no amount of flag waving; no prejudice in favor of some specific

---

SOURCE: "The Scientists' Responsibility" appeared in *Chemical and Engineering News*, April 19, 1971.

regional culture and tradition; no conviction of personal or ethnic superiority, can prevail against the cold equations. The nations of the world must co-operate to seek the possibility of mutual life, or remain separately hostile to face the certainty of mutual death.

Nor can the co-operation be the peevish agreement of haughty equals: each quick to resent slurs, eager to snuff out injustice to itself, and ready to profit at the expense of others. So little time is left and so high have become the stakes, that there no longer remains any profitable way of haggling over details, maneuvering for position, or threatening at every moment to pick up our local marbles and go home.

The international co-operation must take the form of a world government sufficiently effective to make and enforce the necessary decisions, and against which the individual nations would have neither the right nor the power to take up arms.

Tyranny? Yes, of course. Just about the tyranny of Washington over Albany; Albany over New York City; and New York City over me. Though we are each of us personally harried by the financial demands and plagued by the endless orders of the officialdom of three different levels of government, we accept it all, more or less stoically, under the firm conviction that life would be worse otherwise. To accept a fourth level would be a cheap price to pay for keeping our planet viable.

But who on Earth best realizes the serious nature of the problems that beset us? As a class, the scientists, I should think. They can weigh, most accurately and most judiciously, the drain on the world's resources, the effect of global pollution, the dangers to a fragmenting ecology.

And who on Earth might most realistically bear a considerable share of responsibility for the problems that beset us? As a class, the scientists, I should think. Since they gladly accept the credit for lowering the death rate and for industrializing the world, they might with some grace accept a good share of the responsibility for the less than desirable side effects that have accompanied those victories.

And who on Earth might be expected to lead the way in finding solutions to the problems that beset us? As a class, the scientists, I should think. On whom else can we depend for the elaboration of humane systems for limiting population, effective ways of preventing or reversing pollution, elegant methods of cycling resources? All this will clearly depend on steadily increasing scientific knowledge and on steadily increasing the wisdom with which this knowledge is applied.

And who on Earth is most likely to rise above the limitations of national and ethnic prejudice and speak in the name of mankind as a whole? As a class, the scientists, I should think. The nations of the world are divided in culture: in language, in religion, in tastes, in philosophy, in heritage—but wherever science exists at all, it is the same science; and scientists from anywhere and everywhere speak the same language, professionally, and accept the same mode of thought.

Is it not, then, as a class, to the scientists that we must turn to find leaders in the fight for world government?

ISAAC ASIMOV

TODAY AND TOMORROW

One of the greatest science fiction writers takes a fascinating look at scientific fact in today's world – and its implications for the future.

Earth and space, today and tomorrow, hold major challenges for man which he must understand if he is to survive. In this significant volume Asimov tells us what we have learnt so far, what we don't yet know and what we may soon find out. He covers all the vital fields of science to present an astonishing and revealing picture of what is now – and what may be to come.

CORONET BOOKS

ISAAC ASIMOV

ASIMOV ON ASTRONOMY

One of the world's foremost science fiction writers, Isaac Asimov, now leaves the realms of imagination in the even more fascinating search after truth. In this remarkable book, he gives us a highly personal and entertaining insight into the history and science of astronomy, opening up amazing fields of speculation.

Could one of Earth's cities be blown to dust by a killer asteroid?
What is the true feasibility of the Moon becoming an Earth colony?
Is there a tenth planet as yet undiscovered?

Starting from Earth and its neighbours, Asimov takes us on an unforgettable trip through the solar system and the galaxies to encompass, finally, the whole universe.

CORONET BOOKS